MEMS-based Transdermal Drug Delivery

This book introduces transdermal drug delivery and the developments that have taken place in various transdermal drug delivery techniques including the system-level design approach of a novel miniaturized medical device to offer precise and painless drug delivery via a skin-based transdermal route. It discusses the microelectromechanical systems (MEMS)-based fabrication technique and the design, fabrication and characterization of different MEMS-based components like microneedles and micropumps. It further includes a MEMS-based component micropump with design, analysis, fabrication and characterization of the transdermal drug delivery device and the challenges encountered in the design improvements.

Features:

- Summarizes transdermal drug delivery systems especially with a focus on MEMS and microneedles, including theoretical concepts
- Emphasizes system integration by describing simulation and design techniques as well as experimental fabrication
- Discusses system-level integration for miniaturized therapeutic devices
- Includes working simulation models covering microneedles and micropump analysis
- Explores future direction in development of pertinent devices

The book is aimed at researchers, professionals and graduate students in biomedical engineering, microelectronics, micro-electro mechanical systems and drug delivery.

MEMS-based Transdermal Drug Delivery

Richa Mishra and T.K. Bhattacharyya

CRC Press
Taylor & Francis Group
Boca Raton London New York

CRC Press is an imprint of the
Taylor & Francis Group, an **informa** business

Designed cover image: © Shutterstock

First edition published 2024
by CRC Press
2385 NW Executive Center Drive, Suite 320, Boca Raton FL 33431

and by CRC Press
4 Park Square, Milton Park, Abingdon, Oxon, OX14 4RN

CRC Press is an imprint of Taylor & Francis Group, LLC

ISBN: 9781032064239 (hbk)
ISBN: 9781032064246 (pbk)
ISBN: 9781003202264 (ebk)

DOI: 10.1201/9781003202264

Typeset in Times
by Newgen Publishing UK

Contents

About the Authors

Richa Mishra is Assistant Professor in Electronics and Communication Engineering Department, Birla Institute of Technology (BIT), Mesra, India. She pursued her PhD from the Advanced Technology and Development Center, Indian Institute of Technology, Kharagpur and postdoctoral fellowship from the Electronics and Electrical Communication Engineering Department, Indian Institute of Technology, Kharagpur. Her research interests include MEMS, wearable sensors and nano-engineered biosensors. She received the award for the top ten innovations in India in 2016 by the Department of Science & Technology India and Lockheed Martin under India Innovation Growth program (IIGP). She has been working in the area of microneedles and micropumps for last 10 years.

T.K. Bhattacharyya holds the prestigious position of Institute Chair Professor at IIT Kharagpur. Presently, he is the Head of the Advanced Technology Development Centre, IIT Kharagpur, Professor in E&ECE Department and Professor-in-charge Microelectronics and MEMS lab. He has academic and industrial experience of 30+ years. He has published over 250 publications in refereed international journals and 100 in conferences, along with around 11 patents (granted and filed). His prime working area is extensively focused on developing sustainable technologies and technology transfer in the field of MEMS and microsystems, RF and Analog VLSI, nanoelectronics, thin films, nanoscale biosystems engineering, nano-bio sensors, and nano composite-based energy storage and harvesting system.

1 Transdermal Drug Delivery Systems

1.1 INTRODUCTION

Human beings have administered the use of drugs for a long time for therapeutic purposes. The evolution of drug delivery has seen the development in drugs as well as drug delivery devices. In order to differentiate themselves from competitors, the pharmaceuticals and drug delivery companies have turned to innovative system level designs of drug delivery devices. As an example, Bayer's Mirena device is a long used hormonal intrauterine device used for hormonal contraception [1]. GlaxoSmithKline's Breo Ellipta inhaler is a respiratory inhaler employed in asthma treatment. Urogen's delivery system works on the drug targeting the urinary bladder where the drug (Botox+RTGel) in liquid form at low temperatures, converts to gel at body temperatures [2]. It adheres as gel to the bladder wall and provides sustained release of Botox. Novo Nordisk's Flex touch insulin pen targets diabetic patient compliance [3]. AdminPatch microneedle array is a transdermal microneedle array based on pen injector device that injects large quantities of drug across skin when attached with a hypodermic needle [4]. Some of the transdermal drug delivery products already in market (with drugs in adhesive layer) are Catapres-TTS from Alza Corporation, Esclim from Fournier Laboratories, Climara from 3M Drug Delivery Systems, Lipoderm from Endo Pharmaceuticals and duragesic from Alza Corporation [5].

Similarly, there are several ways of drug delivery to skin. These routes derive their names from their starting point from where the drug is administered. Some of these routes are:

a. Buccal
b. Oral
c. Pulmonary
d. Transdermal
e. Ocular
f. Nasal
g. Sublingual
h. Vaginal / Anal

These routes are discussed below in brief.

DOI: 10.1201/9781003202264-1

1.2   DIFFERENT ROUTES OF DRUG DELIVERY

1.2.1   BUCCAL ROUTE OF DRUG ADMINISTRATION

Buccal route has higher total blood flow than the oral route. The buccal mucosa covers the inside of the cheek and lower lips. It has an average surface area of 100 square centimetres. Mucosa present in this region protects the underlying tissues from mechanical and chemical damage. Generally, there are two approaches which define the drug delivery through the mucosa.

    i. Sublingual delivery
    ii. Delivery through buccal mucosa

1.2.1.1   Sublingual Delivery

The sublingual mucosa is a region where rapid onset of drug absorption happens. Hence this route could be used to treat acute pain. But this surface is constantly washed by saliva which presents a problem in full dosage of drug being delivered.

1.2.1.2   Delivery through Buccal Mucosa

Comparatively, buccal mucosa surface is immobile and a preferred site for administration of controlled release systems. Control release systems are one in which drug release from formulation is controlled. For this route of drug delivery, transport of drugs is both transcellular (intracellular) or paracellular (intercellular pathway). Major challenges which are encountered for drug delivery to buccal route are:

    i. How to retain drugs at the application site
    ii. Requirement of enhanced permeability for drugs
    iii. Continuous secretion of saliva
    iv. Chances of swallowing the drugs

1.2.2   ORAL ROUTE

Oral route is the most common route for Drug Administration. This is the most preferred technique of drug administration, but a very scarce quantity of medicine actually reaches our blood after having passed from the body's gastrointestinal tract. In this route, a substance is taken through the mouth and into the gastrointestinal tract. From the gastrointestinal tract, the drugs reached the intestine. Along the way to the intestine, the pill or the capsule is broken down and transported to the bloodstream. In other words, oral route of drug delivery follows the route mouth – stomach – small intestine – colon. It remains a preferred route of drug administration because of the following reasons.

    i. Non-invasiveness
    ii. Patient compliance
    iii. Convenient way of drug administration
    iv. Ease of large-scale manufacturing

Most orally administered drugs are absorbed by the upper parts of the gastrointestinal tract. One would be surprised to note that around 90% of formulations developed for human use are represented by oral formulations [6]. Additional challenges for the oral route of drug administration are the undesirable taste in mouth, gastric degradation, poor aqueous solubility and intestinal mucosal barrier.

1.2.3 INTRAVENOUS INJECTIONS

We are all familiar with the hypodermic syringes that transfer drug directly to blood and it is fastest in being effective. The first disadvantage that is associated with the hypodermic syringes is the needle phobia. The pain and trauma associated with hypodermic syringes decrease its patient acceptability. The other disadvantage is the requirement of a trained staff to administer the injection. This led to exploration of various new techniques associated with the skin route of drug administration. While its advantages include ensuring 100% bioavailability, accurate dose, the disadvantages are that the administration is painful and invasive. Also trained person are needed for administration and there is a widespread needle waste disposal problem.

1.2.4 PULMONARY ROUTE OF DRUG DELIVERY

The respiratory tract is the oldest route used for drug administration. It is a non-invasive drug delivery route. In pulmonary drug delivery, we inhale drug formulation through the mouth. The drug reaches the lungs as it is an absorption as well as treatment organ having large surface area and rich blood supply. The drug enters the bloodstream through alveolar epithelium which are part of alveoli, the sac-like terminals of bronchioles.

The challenges in pulmonary drug delivery include smaller surface area and lower blood flow in the upper airways. Inhaled substances deposit on the mucus covering the conducting airways (comprising the trachea, bronchi and the bronchioles). This region also removes almost 90% of the drug.

Advantages of pulmonary drug delivery include:

i.   Reduced efficiency
ii. Increase surface area
iii. Low drug dose
iv. Rapid onset of action

The technology used by the delivery devices determines the efficiency of the pulmonary drug delivery. Most used devices for pulmonary drug delivery are metered dose inhalers powder dose inhalers and nebulizers.

1.2.5 NASAL ROUTE OF DRUG DELIVERY

In the nasal route of drug administration, the drugs are inhaled through the nose. And Ethan cavity is well vascularized. A drug molecule travels along the cavity across a single epithelial cell layer to be absorbed in systemic blood circulation without first pass hepatic an intestinal metabolism. This is also a non-invasive drug delivery technique. The advantages of nasal drug delivery lake pulmonary drug delivery include lower doses and quicker onset of therapeutic action.

The greatest limitation of nasal route of drug administration is that only selective drugs could be used which are readily absorbed.

1.2.6 OCULAR DRUG DELIVERY

Ocular drugs are delivered locally to the eye. There are multiple paths from precorneal area to systemic circulation from where the drug gets absorbed in the body. These paths are shown below. Although this route of drug administration appears to be easily accessible, the truth is that the eye is well protected from the foreign materials including drugs [7]. The tear flow carries away a portion of drug every time. It is estimated that almost 99% of drug is lost from the precorneal area (Figure 1.1).

The advantages of ocular drug delivery include easy drug administration and quick absorption.

1.2.7 VAGINAL AND ANAL ROUTE OF DRUG DELIVERY

The vaginal route of drug administration is important from both local and systemic diseases point of view [8]. While the vagina is the entry point for drugs in the vaginal

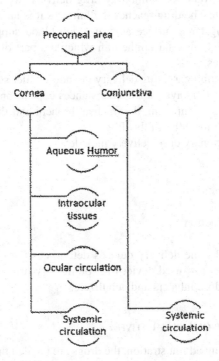

FIGURE 1.1 Ocular drug delivery routes.

route, the anal or rectal route of drug administration uses the rectum as the drug delivery route where drugs are absorbed by the rectum's rich blood vessels and transported to systemic circulation of body. The advantages of these routes include large surface area, rich blood supply and avoidance of first pass hepatic metabolism. The challenges of vaginal and anal routes of drug delivery include the mucosal barriers and potential degradation of drugs and low patient compliance.

1.2.8 Transdermal Route of Drug Administration

Another drug delivery technique which is only next to oral drug delivery in popularity is the transdermal route of drug delivery. The challenge encountered in transdermal drug delivery is that the stratum corneum, the topmost skin layer, is a barrier to drug diffusion across skin. Also, the intercellular space has a lipidic nature and makes transport of polar or hydrophilic molecules very difficult. Transdermal routes of drug delivery include both invasive and non-invasive methods. One of the main advantages of the transdermal drug delivery system is that it is administered through skin, so it passes the gastrointestinal tract and suffers no loss due to first pass metabolism. Secondly, in transdermal drug delivery the bioavailability of drugs increases.

As we saw that the major limitations encountered by the transdermal drug delivery systems is posed by the physical barrier of skin itself. Skin is extremely flexible and hence poses a mechanical resistance to any external object trying to enter skin. The mechanical and structural properties of skin vary widely between individuals, skin types, hydration, body location and age [9]. It is extremely important to understand the skin anatomy if we want to understand and design transdermal drug delivery systems. We shall now look at the skin structure and transdermal route of drug delivery in detail [10].

1.3 SKIN

Skin can be divided into three main layers: Epidermis, dermis and hypodermis (shown in Figure 1.2). The epidermis (50–150 μm thick) consists of outward moving cells which are constantly renewing. The outermost layer of the skin (10–20 μm) is called the stratum corneum. This thick layer of dead cornified hexagonal shaped cells also acts as a water barrier to skin. Next is the dermis (1.1 mm thick) which is the bulk of the skin. It gives tensile strength and elasticity to skin and support for nerves and the vasculature network. Below the dermis is the hypodermis which is around 1.2 mm thick. It consists of loose fatty tissues which vary widely between individuals [11].

1.3.1 Epidermis

It is the topmost layer of skin which is 75–150 μm in thickness. The epidermis could be viewed to be made up of different layers having their characteristic components.

   a. Stratum corneum
   b. Granular layer

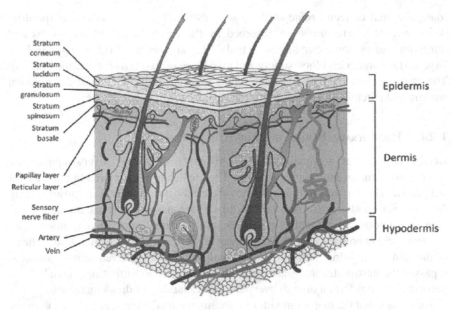

FIGURE 1.2   Different layers of skin.

   c. Spinous layer
   d. Basal layer

1.3.1.1   Stratum Corneum

It is the outermost layer of skin. Around 15–30 layers of cells in the stratum corneum are made up of protective cells. They prevent external bacteria, fungi and viruses from reaching the inner layer of skin. They also provide protection to inner layers of skin from friction and abrasion. It also prevents water evaporation. An interesting process called desquamantion takes place in this layer as the new cells travel from the basal layer and keep replacing the stratum corneum in about four weeks. **Stratum Lucidium** – Lucidium refers to the clear look of the cells in this layer. This region exists only in the thick skin on palms and soles of feet.

1.3.1.2   Granular Layer

As the name suggests, this layer is packed up of tiny granules produced in skin cells. Keratin which provides strength and waterproofness to skin is packaged in these granules along with moisture. These granules break to release their content in inter-cellular spaces and transport them to the stratum corneum.

1.3.1.3   Spiny Layer

The main purpose of this layer is to provide strength to epidermis. The cells in this layer have tiny protrusions which adhere the cells together.

1.3.1.4 Basal Layer

The basal layer is the deepest layer of the epidermis. Keratinocytes are produced in this layer where they move to the outermost layer and replace the dead cells. This layer has Merkel cells and melanocytes. Merkel cells account for your sense of touch. Melanocytes produce melanin which imparts colour to our hair and skin [12].

1.3.2 Dermis

Dermis is the second layer of skin beneath epidermis which is 1.5–2.5 mm in thickness. It provides support to epidermis. It contains tissues and a lot of blood vessels. The main components of dermis include collagen (structural protein), elastin (a stretchy protein), a clear gel-like fluid filling the intercellular space, sweat glands and hair follicles. Dermis mainly consists of two layers – papillary dermis and reticular dermis.

1.3.2.1 Papillary Dermis

This is the top layer of the dermis and lies below the epidermis forming a strong interlocking. It consists of fat cells, blood vessels, collagen fibres, nerve fibers and touch receptors.

1.3.2.2 Reticular Dermis

It is the bottom layer of the dermis which lies below the papillary dermis. It is thicker than papillary dermis and made up of hair follicles, nerves, blood vessels and fat cells. Elastin fibres and collagen fibres make a net-like structure surrounding this layer. This net like structure accounts for providing support to the entire skin and at the same time allowing it to move and stretch. The nerve ending in the dermis allows us to feel different sensations like pain, warmth, cold touch and irritation. This layer also has sweat glands that maintain body temperature.

1.3.2.3 Hypodermis

Hypodermis is the bottom layer of skin. This layer connects the dermis layer to the muscles and bones and protects them from harm. It is also known as the subcutaneous layer. Many injections like epinephrine for allergic reactions, insulin and anti-arthritis drugs are delivered to this layer. This layer consists of fibroblasts (main constituent of connective tissues), adipose tissues, macrophages (type of white blood cells) and connective tissues holding nerves and blood vessels together.

1.4 DRUG DELIVERY ROUTES IN SKIN

There are a number of ways by which external chemicals could cross the stratum corneum [13]. They are:

    a. Intercellular
    b. Intracellular
    c. Transfollicular
    d. Mechanical delivery methods

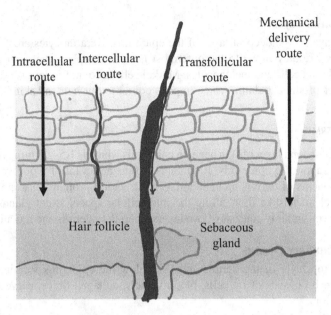

FIGURE 1.3 Schematic pathway of different routes of drug delivery in "brick and mortar" skin model (Corneocytes (bricks) are separated by the lipid matrix (mortar)).

The different routes of transdermal drug delivery are shown in Figure 1.3.

The intercellular route, also known as extracellular route, consists of the lipid matrix. The densely packed extracellular lipid matrix is a major skin barrier for drug delivery. It is made up of free fatty acids (lipid species released from several cell types), cholesterol and ceramides. There are two dominant lamellar phases for this area. A lamellar phase is formed when the headgroups of the lipids face the aqueous phase on both sides of the bilayer with the hydrocarbon chains opposing each other. Intracellular route or transcellular route is the chemical absorption in skin via the cells. The intracellular is contributed by both corneocytes or lipid matrix. Mechanical methods to breach the stratum corneum include various methods like sonophoresis, microneedles, thermal ablation, iontophoresis etc.

1.5   MECHANISM OF PAIN SENSATION

Pain sensation is an important management part when it comes to any technique of drug delivery. Hence, it is important to understand the pain pathway that leads to pain sensation. Figure 1.4 shows the mechanism of pain sensation.

When something sharp pricks our fingers, it causes tissue damage. This damage is picked up by microscopic pain receptors called nociceptors in skin. These pain receptors are one end of a nerve. The axon in the spinal cord connects one nerve end to another. Once the pain receptor is activated, it sends electrical signal to other nerve end. The nerves communicate with each other through neurotransmitters and electrical signals related to the event of tissue damage reaches the spinal cord in

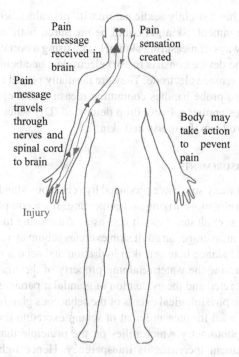

Pain message received in brain

Pain sensation created

Pain message travels through nerves and spinal cord to brain

Body may take action to pevent pain

Injury

FIGURE 1.4 Mechanism of pain sensation in brain.

the neck. Signals are then passed from spinal cord to brain and leads to pain sensation. Different sensations are carried by different type of nerve fibres. Common nerve fibres are A-alpha, A-beta, A-delta and C nerve fibres. Sharp and pricking pain is carried by A-delta fibres while dull and throbbing pain travels through C nerve fibres. A-delta fibres conduct faster than C fibres as they are electrically insulated. Thus, on pricking one's fingers, sharp pain is felt accompanied by a slow throbbing pain.

1.6 MEASUREMENT OF SKIN PARAMETERS

The outermost layers of skin present a major barrier in transdermal drug delivery. Hence it is important to measure the skin barrier. This skin barrier also regulates the drug flow through skin. Measurement of multiple parameters helps in comprehensive analysis of skin. Few skin parameters and their measurement techniques are given below.

1.6.1 Skin Surface pH

Previous sections gave us an idea of the different layers of skin. We had seen that the topmost layer of skin consists of dead cornified cells. Our body displays a wonderful balance on this top layer by a process called homeostasis where granular layer keratinocytes are converted to corneocytes and then eliminated as dead cells at the skin surface. An important regulation which maintains this balance is the skin surface pH. The skin surface has a slightly acidic nature with pH values between 4–6 than the body's internal environment. Skin pH could be measured both invasively and non-invasive. One of the ways of measuring skin pH is by using a portable pH meter using planar electrodes. The device consists of two electrodes, one being a sensitive electrode and the other reference electrode. They are normally placed in a single housing. They are connected to probe handles containing measurement electronics. Skin pH measurement includes applying 1 or 2 drop deionized (DI) water or saline and then placing the flat electrode on the moistened skin.

1.6.2 Sebum Measurements

Sebum is an oily and waxy substance produced by sebaceous glands in dermis. It acts as a skin coating thereby moisturizing and protecting. It's a complex mixture of free fatty acids, glycerides, cholesterol etc. It acts as a skin coating thereby moisturizing and protecting it. On an average, an adult skin excretes sebum at a rate of 1 mg/10 cm^2 every 3 hours. Right balance between skin hydration and sebum results in a healthy skin. They also determine the water retaining property of the stratum corneum. The amount of sebum excreted and the evaluation of glandular parameters give important information about the physiological events of the sebaceous glands. A device called a sebumeter is widely used for measurement of sebum excretion levels on skin. Often it uses grease spot photometry which relies on the principle that sebum deposited on a translucent element increases its transparency. Hence light passing through this area could be measured using a photoelectric receiver. Based on the above principle, various techniques like UV visible spectrophotometry and other photoelectric techniques provide information about sebum droplet size and distribution.

1.6.3 Skin Thickness Measurement

The different layers of skin vary over the body. It also varies over age, gender, race and ethnicity. In practice, skin thickness is important with respect to assessment of nutritional and dietary information of an individual. Many times, skin thickness could also predict progression of diseases like Type 2 diabetes [14]. Several in-vivo techniques of measuring skin thickness are mentioned below.

   i.   Skinfold Caliper – it is used to measure skinfold thickness so that prediction of total amount of body fat could be done.
  ii.   Computerized Tomography – Computerized Tomography is an imaging procedure which uses X-Ray to obtain internal image body parts. Thin cross-sectional radiography of skin indicates presence of fat in abdominal tissues.
 iii.   Ultrasound Measurement – Epidermal and dermal thickness can be estimated using high frequency (22–55 MHz) ultrasound imaging.

1.6.4 Trans Epidermal Water Loss (TEWL)

As already discussed, skin barrier is important for a body. Internally it maintains homeostasis and protects skin from water loss. Externally, it protects skin from external agents like allergies and chemicals. Skin barrier function is characterized by a technique which measures trans epidermal water loss. This water loss is the amount of water that generally evaporates through skin to the outside environment due to pressure gradient on both sides of skin. This water loss amounts to around 300–400 mL/day. This technique indicates the corneocytes (basic blocks of stratum corneum) availability in skin. Lesser the number of corneocytes, higher are the TEWL values. For TEWL measurements, the water which evaporates from skin is measured with the help of a probe placed in contact with skin [15]. It consists of sensors which detect change in water vapor density. TEWL can be measured by following devices.

i.   Open chamber device – A hollow device placed in contact with skin for calculating humidity gradient.
ii. An unventilated chamber device – A closed upper end chamber which protects measurement from ambient condition.
iii. Condenser chamber device – provides highest sensitivity among all three techniques. It provides dynamic reading of transcutaneous water loss.

1.6.5 Young's Modulus Measurement

Young's modulus is one of the vital parameters to estimate skin characteristics. It yields important information about mechanical behaviour of skin which is required for many applications. We know that as skin is subjected to stress, initially a low stiffness is observed due to the shifting of the wavy collagen bundle fibres. When this force is increased, it then yields the characteristic load displacement curve. Different techniques which give an idea about skin's Young's modulus are tensile or indentation tests. Skin may be stretched (tensile test), normal load may be applied to skin (indentation test), elevated in an aperture (suction test) or rotated (torsion test) to give mechanical behaviour details [16].

1.7 WAYS OF TRANSDERMAL DRUG DELIVERY

Transdermal drug delivery provides a skin route of drug delivery where the drug enters through the stratum corneum and reaches the underlying epidermis and dermis layer and then gets absorbed in the systemic circulation to yield steady blood circulation. Over the years, various transdermal drug delivery techniques have been developed. Some of them are discussed below (Figure 1.5).

1.7.1 Sonophoresis

In Sonophoresis, ultrasound is used to enhance absorption of drugs through the stratum corneum. This technique typically uses a low frequency wave (20 kHz-16 MHz) to increase the kinetic energy of molecules which enable them to disrupt the stratum corneum lipid layer. After disruption the drug molecules can easily permeate the skin [17]. During sonophoresis, following phenomena are known to take place,

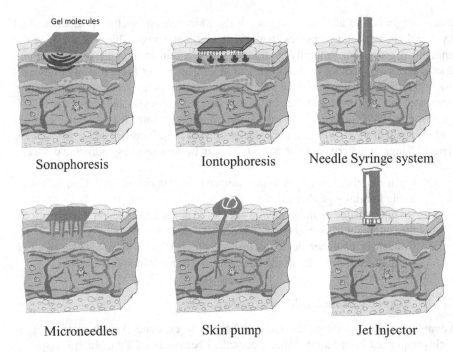

FIGURE 1.5   Various ways of transdermal drug delivery.

(i)   Acoustic streaming
(ii) Cavitation
(iii) Thermal effect

In acoustic streaming, a steady state velocity is generated by absorption of ultrasonic waves by viscous fluid. Then in the next phase of sonophoresis, cavitation takes place. The application of ultrasound leads to generation of small bubbles which lead to cavity generation due to implosion near the stratum corneum. This introduces defects in the skin lipid layer. Cavitation leads to breaking of cell membranes in top layer of skin thus increasing skin permeability. Wong et al. 2014 [18] showed that these cavitation bubbles result in micro jets targeted towards skin. This may create water pathways through the lipid layer through which drug can be transferred. Thermal effect comes from the heat generated by sonophoresis. This local heating of skin results in making the skin more porous and elevating the kinetic energy of the drug molecules. The disadvantages of this technique are that it is not recommended in case of cuts. It may lead to minor irritation and may require larger time for drug delivery.

1.7.2 IONTOPHORESIS

Interestingly iontophoresis is made up of two words "ionto" and "phoresis". "Ionto" refers to positively or negatively charged ions and "phoresis" describes ion transport across skin. It is a non-invasive method to enhance drug permission across skin. In iontophoresis, drug particles (precisely ions of soluble salts) move through a fluid into the body tissue. During the process, mild electrical currents are produced to deliver medication across skin. The ions prefer to move across skin through the route of least resistance. The sweat gland pores are one such route [19].

The drugs which are commonly delivered using iontophoresis are opioids, steroids, antibacterial, antifungal drugs and vitamin E [20]. The iontophoresis-based drug delivery device consists of the following components [21].

(i) Power supply
(ii) Electrode
(iii) Control circuit
(iv) Drug storehouse
(v) Electrolyte storehouse

In order to drive the ions in skin, a current is required. The drug to be delivered should be of such formulation that it is ionic in nature. The positively charged ions should be placed under a positively charged electrode. Similar arrangements should be made for negatively charged ions. The basic principle is that the electrodes will repel the similarly charged ions in the skin. Out of the two electrodes used, the electrode with the ionic drug solution is called the active electrode while the other electrode is called inactive electrode which is simply used to complete the circuit. In iontophoresis, the drug ions are delivered in skin through skin pores like sweat glands and hair follicles instead of stratum corneum. This is because the stratum corneum has higher resistance compared to pores. This makes the pores a preferred route for ions [22]. The advantages of iontophoresis include:

(i) Easy administration and enhance penetration.
(ii) Non-invasive.
(iii) Continuous or pulsatile drug delivery could be achieved using iontophoresis.
(iv) Aids in delivery of large molecular weight as well as polar molecules.

The disadvantages include risk of skin burn if the electrodes are not handled properly. Also, this method calls for a lot of optimization in order to stabilize the therapeutic agent.

1.7.3 Needle Syringe System

The hypodermic syringes transfer drug directly to blood and it is fastest in being effective. The first disadvantage that is associated with the hypodermic syringes is the needle phobia. The pain and trauma associated with hypodermic syringes decrease its patient acceptability. The other disadvantage is the requirement of a trained staff to administer the injection.

1.7.4 Microneedles

Microneedles are needles having dimensions in the micrometre range and they deliver drug to body by puncturing the outermost layer of skin (stratum corneum). The drug reaches the veins either directly or through diffusion and is then absorbed in the body through systemic circulation. While the microneedle is inserted inside the skin, it has to overcome the barrier caused by the outermost layer of the skin [23, 24]. The skin puncturing force falls drastically once the skin is punctured. The insertion by microneedles is perceived as painless and they are self-administrable [25, 26].

1.7.5 SKIN PUMPS

Skin pump-based drug delivery aims to achieve the therapeutic effect for a variety of drugs. They are used to pump the drugs which find it difficult otherwise to penetrate the body. [27–29]. Various actuations mechanisms used in the pumps are diaphragm displacement, piezoelectric, electrostatic, electromagnetic, pneumatic and thermopneumatic actuation. Their advantages lie in the precise control of the drug flow rate (where they are human compliant either at fast or slow rates) and extremely small size [30, 31]. Skin pumps are attached to the body with the help of catheters.

1.8 CONCLUSION

There are various routes of drug delivery to the human body. Each of these routes present their own advantages and disadvantages. Over the time, focus on drug delivery platform development has shifted to patient compliance, larger bioavailability of drugs and self-administration techniques. After having discussed these techniques in brief, the chapter shifts towards the transdermal drug delivery route. This route has gained quite popularity in recent years due to the factors mentioned above. The major barrier to transdermal routes of drug delivery is skin itself. Hence it becomes important to understand skin parameters and their measurement techniques. The transdermal drug delivery technique has also seen development through various ways which have also been discussed in the chapter. Thus, the ground has been created for the readers for which drug delivery device design and fabrication shall be looked into.

REFERENCES

1. J. Agmon, E. Soronker, Intrauterine device and inserter for the same, US patent, US8573222B2, 2010.
2. A. Holzer, D. Daniel, E. Hirszowicz, Methods for treatment of bladder cancer, US patent, US9011411B2, 2004.
3. C.P. Enggaard, C.S. Moller, Fredensborg, T. HedeMarkussen, Dose mechanism for an injection device for limiting a dose setting corresponding to the amount of medicament left, US patent, US9775953, 2017.
4. V.V. Yuzhakov, The admin pen microneedle device for painless & convenient drug delivery, *Drug Deliv. Technol.* 10(4):32–36, 2010.
5. T. Tanner, R. Marks, Delivering drugs by the transdermal route: review and comment, *Skin Res. Technol.* 14(3):249–260, 2008.
6. S. Alqahtani Mohammed, K. Mohsin, A. Alsenaidy Mohammad, Z. Ahmad Muhammad, Advances in oral drug delivery, *Frontiers Pharmacol.* 12, 2021. DOI: 10.3389/fphar.2021.618411
7. C. Bucolo, F. Drago, S. Salomone, Ocular drug delivery: a clue from nanotechnology, *Frontiers Pharmacol.* 3, 2012. DOI: 10.3389/fphar.2012.00188

8. F. Acartürk, Mucoadhesive vaginal drug delivery systems. *Recent Pat. Drug Deliv. Formul.* 3(3):193–205, 2009. DOI: 10.2174/187221109789105658. PMID: 19925443

9. L.P.O. Norten, *The Skin Barrier Structure and Physical Function.* Ph.D. Dissertation, Karolinska Institute, Stockholm, Sweden, 1999.

10. W.Y. Jeong, M. Kwon, H.E. Choi, K.S. Kim, Recent advances in transdermal drug delivery systems: A review, *Biomater. Res.* 25(24), 2021. https://doi.org/10.1186/s40824-021-00226-6

11. R.C. Haut, *Biomechanics of Soft Tissue*, 2nd edition, New York: Springer, 28–53, 2002.

12. H. Joodaki, M.B. Panzer, Skin mechanical properties and modeling: A review, *Proc. Inst. Mech. Eng., Part H: J. Eng. Med.* 232(4):323–343, 2018. DOI: 10.1177/0954411918759801

13. G. Honari, H. Maibach, eds., Skin structure and function, *Applied Dermatotoxicology*, Academic Press, 1–10, 2014.

14. A.J. Alkhatib, A. Sindiani, K Funjan, E. Alshdaifat, Skin thickness can predict the progress of diabetes type 2: A new medical hypothesis, *EC Diabetes Metab. Res.* 4(8):8–12, 2020.

15. H. Alexander, S. Brown, S. Danby, C. Flohr, Research techniques made simple: Transepidermal water loss measurement as a research tool, *J. Invest. Dermatol.* Nov 138(11):2295–2300, 2018. DOI: 10.1016/j.jid.2018.09.001. PMID: 30348333

16. A. Kalra, A. Lowe, A.M. Al-Jumaily, Mechanical behaviour of skin: A review, *J. Mater. Sci. Eng.* 5(4), 2016. DOI:10.4172/2169-0022.1000254

17. K. Ita, ed., Sonophoresis, *Transdermal Drug Delivery*, Academic Press, 231–255, 2020. ISBN 9780128225509, https://doi.org/10.1016/B978-0-12-822550-9.00010-7

18. T.W. Wong, Electrical, magnetic, photomechanical and cavitational waves to overcome skin barrier for transdermal drug delivery, *J. Control Rel.* 193:257–269, 2014.

19. S. Rawat, S. Vengurlekar, B. Rakesh, S. Jain, G. Srikarti, Transdermal delivery by iontophoresis, *Indian J. Pharm. Sci.* 70(1):5–10, 2008. DOI:10.4103/0250-474X.40324

20. T.M. Karpiński, Selected medicines used in iontophoresis, *Pharmaceutics.* 10(4):204, 2018. DOI: 10.3390/pharmaceutics10040204

21. N. Kanikkannan, Iontophoresis-based transdermal delivery systems, *BioDrugs.* 16:339–347, 2002.

22. W. Yeup Jeong, M. Kwon, H. Choi, K.S. Kim, Recent advances in transdermal drug delivery systems: A review, *Biomater. Res.* 25(24), 2021. https://doi.org/10.1186/s40824-021-00226-6

23. T. Miyano, Y. Tobinaga, T. Kanno, Y. Matsuzaki, H. Takeda, M. Wakui, K. Hanada, Sugar micro needles as transdermic drug delivery system, *Biomed. Microdevices.* 7:185–188, 2005.

24. D. Wermeling, S. Banks, D. Hudson, H. Gill, J. Gupta, M. Prausnitz, A. Stinchcomb, Microneedles permit transdermal delivery of a skin impermeant medication to humans, *Proc. Natl. Acad. Sci.* 105(6):2058–2063, 2008.

25. A. Kaushik, A.H. Hord, D.D. Denson, D.V. McAllister, S. Smitra, M.G. Allen, M.R. Prausnitz, Lack of pain associated with microfabricated microneedles, *Anesth. Analg.* 92:502–504, 2001.

26. R.K. Sivamani, B. Stoeber, G.C. Wu, H. Zhai, D. Liepmann, H. Maibach, Clinical microneedle injection of methyl nicotinate: stratum corneum penetration, *Skin Res. Tech.* 11(11):152–156, 2005.

27. H.A.E. Benson, A.C. Watkinson, eds., *Topical and transdermal drug delivery: principles and practice*, 6–7, Wiley, 2012, ISBN 978-0-470-45029-1

28. E.A. Tetteh, M.A. Boatemaa, E.O. Martinson, A review of various actuation methods in micropumps for drug delivery applications, *2014 11th International Conference on Electronics, Computer and Computation (ICECCO)*, Abuja, 1–4,2014.

29. C. Joshitha, B.S. Sreeja, S. Radha, A review on micropumps for drug delivery system, *2017 International Conference on Wireless Communications, Signal Processing and Networking (WiSPNET)*, Chennai, 186–190, 2014.

30. P.K. Das, A.B.M.T. Hasan, Mechanical micropumps and their applications: A review, *AIP Conf. Proc.* 1851(1):020110, 2017.

31. Y.N, Wang, L-M Fu, Micropumps and biomedical applications – A review, *Microelectron. Eng.* 195:21–138, 2018.

2 At the Bottom
Microfluidics and Scaling Laws

2.1 INTRODUCTION

How often does the handling and manipulating of substances at micro size, like sugar crystals, frustrate you? I want to suggest the possibility of training an ant to train a mite to do this. What are the chances of small but movable machines? The manufacturing process that creates tiny devices was developed back in 1983. These small devices are called Micro-electromechanical Systems (MEMS) that have a length of 1 micron which combine electrical and mechanical components. Recent fabrication techniques for MEMS include silicon micromachining, photolithography, 3D printing, etc. Interest in Bio-MEMS has increased and found widespread application in biomedical engineering, diagnostics, drug delivery, etc. [1]. Micro-pumps are used for ink jet printing. Potential medical applications for small pumps include controlled drug delivery and monitoring the minute amount of medication [2]. Flow control using MEMS promises a jump in the control system. The ability to create structures on micron length has brought a wide range of scientific investigations and devices to transport fluids and pattern surfaces. These types of investigation involving fluids are defined under microfluidics.

2.2 MEMS APPLICATIONS

MEMS technologies can provide a means to the digital world that is dominated by Integrated Circuits. Sometimes, it is advantageous to link transduction mechanisms in series. Moreover, the sensing and actuating mechanisms can be combined with electronics to form complete micro-systems.

2.2.1 ADVANTAGES OF MEMS

- Good Scaling properties: Performs better when miniaturized to micrometer scale.
- Batch Fabrication: Cost of producing one MEMS device is nearly equal to the cost of producing many MEMS devices.

DOI: 10.1201/9781003202264-2

2.3 MICROFLUIDICS

Microfluidics is the concept of fluids and submillimeter-sized systems influenced by external forces. So, small-scale microfluidic channels and fluidic behaviour deviates from microfluidic behaviour. This technology of fluid manipulation can be described by fundamental equations used to study fluid physics in the macro-scale. Therefore, microfluidics can be categorized under Fluid Mechanics. The channel's height or width (or both) in microfluidic systems is a few micrometers. At the same time, it can hold the volume of liquid in the field of milliliters (10^{-3}) to nanoliters (10^{-9}) or smaller. In micro-scale, diffusive mass transport takes control over convective mass transport, given by **Sherwood Number (S_h)**, and is defined as [3].

$$S_h = \frac{k*d}{D} \tag{2.1}$$

'D' is diffusion coefficient, 'k' is mass transport coefficient, 'd' is characteristics diameter of the channel. For macro-scale, S_h is large, which indicates that convective transport takes control over diffusive mass transport.

Microfluidics is a small platform with channel systems connected to reservoirs where the channel size is in micrometers. The concept of microfluidics was keen to significantly decrease sample consumption and increase efficiency in the separation method [4].

Fabrication of microfluidics requires cleanroom facilities and specific equipment, therefore the majority of these devices are simply fabricated using photolithography on substrates such as silicon, etc. [5]. Microfluidics have a relationship with the derivation of Hagen-Poiseuille's Equation; therefore, later, Poiseuille's published it, and that equation is called '**Poiseuille's Equation,**' which means there is a relationship between the resistance arising during fluid flow and the channel length [6].

$$\Delta p = -\frac{8\mu L}{\pi r^4} Q \tag{2.2}$$

The actual work that drove the development of microfluidics was in 1974 at Stanford by Stephen Terry. The author designed a miniaturized gas chromatography (GC) unit and registered it as the first "Laboratory-on-a-Chip" (LOC) device [7].

2.4 IMPORTANCE OF MICROFLUIDICS

   i. *Efficient:* The role of microfluidics for point-of-care testing (POC) devices is crucial, the possibility of microscopy that requires a low Reynolds number, is a laminar flow. As a result, there is a reduction in the required reagent volume and sample size.

   ii. *Reliability:* There is also a possibility of carrying out multiple analyses simultaneously because of the ease of containing various micro-channels within a single device. Detection effectiveness was achieved by tuning the channel design, architecture, and geometry. As a result, the processing time compared to conventional diagnostic methods reduces to a few minutes from several hours.

iii. *Portability:* Often smaller than palm size, microfluidics devices carry some inherent advantages, namely accessibility, portability, and ease of handling/usage.

iv. *Cost-effective*: Besides being efficient, the fabrication cost of microfluidics-based diagnostic devices is much lower than conventional diagnostic techniques. In conventional methods, the total cost of the aspects mentioned above adds up to a considerable amount. While polymeric materials like PDMS, Polymethylmethacrylate (PMMA), polystyrene (PS), and polycarbonate (PC) are utilized for the production of devices that are not only cost-effective but can also be quickly processed. Paper is also one such cost-effective material that can be used for the fabrication of such microfluidic devices that are lightweight, disposable, and biocompatible.

2.4.1 APPLICATIONS OF MICROFLUIDICS

i. **Flow chemistry:** Microfluidics is linked to the synthesis of materials, where the reactions governing the synthesis process are carried out inside a micro-channel. This synthesis platform opens up opportunities for the industrial-scale production of materials.

ii. **Analytical devices:** These types of microfluidic devices are used to mimic bulky columns for chromatography and mass spectroscopy. Such microfluidic devices utilize relatively low sample concentrations and volumes and provide analytical results faster.

iii. **Drug delivery:** Microfluidics-based devices are also used in invasive drug delivery applications in which microneedles, micro-pumps, and inhalers are used to precisely deliver small volumes of a drug at specific target sites.

iv. **Pharmaceutical research:** In pharmaceutics, microfluidics devices are used to discover and screen new drugs. Cell analysis is also performed within micro-channels with high accuracy.

v. **Point-of-care diagnostic devices:** Such microfluidic devices are used for diagnostics at places away from the laboratory or locations where complex systems cannot be installed. POC diagnostic devices are a viable tool for testing at home or in remote areas without trained personnel.

2.5 LOW REYNOLDS NUMBER IN MICROFLUIDICS

The channels fabricated till date have dimensions l=1–300 μm and flow speeds 'u,' and for liquids, they are in the range up to cm/s which yields a Reynolds number, $R=\rho u l/\mu < 30$.When, $R<1$, viscous forces dominate. Secondly, when the Reynolds Number is not less than 1, then the flow becomes laminar [8].

Therefore, **Laminar flow** is another property of microfluidic systems shown by the **Reynolds Number (Re),** defined as: [9]

$$R_e = \frac{\rho \vartheta d}{\eta} \qquad (2.3)$$

Where, η is dynamic viscosity of the fluid, ρ is mass density of the liquid, υ is fluid velocity. Due to the micrometer size of microfluidics systems, the Reynolds Number can be less than 1, i.e. viscous forces rules over the flow, and is said to be laminar. The origin of microfluidics is from microelectronics technology, where fabrication technique used is Photolithography. The foundational work in microfluidics used Silicon as the material. This is used when chemical stability is required for biological assays. Therefore, the following efforts towards microscopic scale is to decrease the time analysis and better performance, the concept of LOC appeared [10]. Microfluidics' central focus is building, designing, and operating systems that can employ drugs and materials in miniaturized fluid volumes. These devices are called "Lab-on-a-Chip" systems.

2.6 LAB-ON-A-CHIP (LOC)

LOC is a microfluidic platform approached by scientists from various domains and has many applications. A linear reduction by a factor of 10^3 amounts to a volume reduction by a factor of 10^9, Therefore, instead of handling 1 L, a lab-on-a-chip system can deal with 1 nL; thus, such a small amount allows for quick and efficient analysis [11]. LOC systems can be thought of as general ICs and Microelectromechanical systems (MEMS).

Lab-on-a-Chip (LOC) devices are multi-functional as shown in Figure 2.1 on a single chip. This micro fluidic platform provides automated, and parallelized chemical and biological analysis [12] that offers low-cost, faster, controllable biological assays at a microscopic level. These micro-engineered devices can handle minimal fluid volumes down to less than a few Pico-liters [13]. Micro-droplets of whole blood, saliva have the capability to get tested in these mini platforms for diagnosis [14]. One drop of the patient's blood is required to supplant on the inlets of the device, then the entire test takes place on a single platform, within minutes [15]. Other biomedical applications for LOCs have been reported, including proteins and DNA detection, etc.

2.7 APPLICATION OF LAB-ON-A-CHIP SYSTEMS FOR POINT-OF-CARE TESTING (POCT)

Point-of-care testing systems diagnostic platforms are micro biomedical devices which can provide test results immediately in the easiest way [16]. These devices do not need trained practitioners, as the patients are able to do tests at home. The need for immediate diagnosis of acute diseases has grown the interest in developing the micro POCT systems. Since, the large surface area to volume ratio in microfluidic systems decreases the time of test in LOC for POC testing [17].

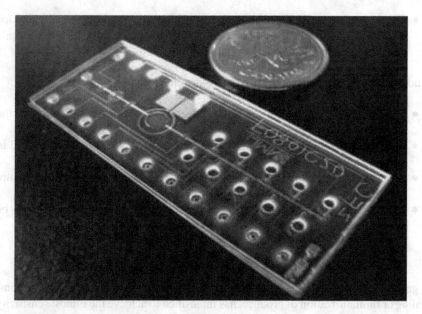

FIGURE 2.1 Lab-on-a-chip device.

These diagnostic devices are cost-effective, easy to handle, tools to detect different biomarkers, including proteins. Therefore, fast detection of POCT devices helps to increase patients' survival rate.

Adam Heller et al. [18] found that electrochemical glucose monitoring has massively improved the lives of diabetic people. The single test strip can accurately and painlessly monitor glucose in 300 nL samples of blood. Liqiang Liu et al. [19] developed a lateral immuno-chromatographic assay for detecting 19-nortestosterone in which the testing time was less than 20 min, therefore, it was suitable for rapid on-site tests.

Therefore, LOC aims to obtain results near the patient's location to plan the treatment [20]. Microfluidics are used for laboratory experiments, such as drug testing, [21–30], point-of-care diagnosis [30,31], These chips are made up of PDMS, which is a commonly used transparent elastic polymer [32]. It is a standard material that can be used to fabricate the microfluidic devices (MFDs), which obeys Henry's law [33].

Therefore, LOC devices require different techniques to achieve passive operations.

- In microfluidics, the generation of submissive flow can be obtained from appropriate surface and liquid combinations that can create the required solid–liquid surface tension gradients [34]. Du et al. reported that the concentration gradient in the channel induced forward flow and was enhanced by the evaporation-induced backward flow. Due to this, the gradient generated was controlled by convection and molecular diffusion [35].

- Fluid pressure means force per unit area. Pressure in liquids is equally divided in all directions; so, if force is applied to one point of the liquid, it will be transmitted to all other issues in the fluid [36].

Fluid motions in these micron sized systems are characterized by-

- Application of pressure difference – parabolic velocity distribution across the channel.
- Electric fields – electro-osmotic flow when driven by stresses concentrated in charged layers.
- Capillary driving forces owes to wetting of surfaces which leads to pressure gradients in liquids.
- Free surface flow driven by gradient in interfacial tension and these can be manipulated using the dependence of surface tension on temperature.

2.8   SCALING LAWS AND GOVERNING EQUATIONS

In general, scaling means a reduction in the size of a system from all directions. Scaling or miniaturization of products has intensified to achieve the utmost sensitivity and robustness but is cost-effective. Generally, these scaling laws are the relationship between dependent and independent variables. In microfluidics, if L is the linear dimension, volume ($V = L^3$) and surface ($SA = L^2$), then, the scaling law of system dimensions is:

$$\frac{SA}{V} \propto L^{-1} \tag{2.4}$$

Hence, surface forces dominate over the body forces in a microfluidic system.

In fluid mechanics, two quantities, volumetric flow (Q) and pressure drop (ΔP) is considered to be necessary. From Equation 2, Q, the volumetric fluid flow rate is directly proportional to the fourth power of radius r.

$$\mathbf{Q} \propto r^4 \tag{2.5}$$

This equation implies that if the radius is reduced by ten times, the volumetric flow is reduced by 104 times. Since, from Equation 2:

$$\frac{\Delta P}{L} \propto r^{-3} \tag{2.6}$$

With a reduction of the radius by ten times, there is an increase in the pressure drop by a value of 103 times per unit length.

2.9 SCALING OF DIMENSIONLESS NUMBER WITH LENGTH SCALE IN MICROFLUIDICS

With the understanding of dimensionless numbers in microfluidics, there is a reduction in the variables used to describe a given system. There are various types of dimensionless numbers in microfluidics:

2.9.1 REYNOLDS NUMBER

The concept of the Reynolds Number was introduced in 1851 by George Stokes and popularized in 1883 by Osborne Reynolds given in Table 2.1 and Figure 2.2. It is defined as the ratio of internal forces to the viscous forces.

$$\mathbf{R}_e = \frac{inertial\ force}{viscous\ force} = \frac{\rho \vartheta D}{\mu} = \frac{\mu D}{\vartheta} \tag{2.7}$$

If we want to scale the Reynolds number with 'L':
  When velocity (u) is constant:

$$R_e = \frac{\rho \vartheta D}{\mu} \propto L \tag{2.8}$$

  When pressure difference (Δp) is constant:

$$R_e = \frac{\rho \vartheta D}{\mu} \propto L^2 \tag{2.9}$$

2.9.2 KNUDSEN NUMBER

It is the ratio of the mean free path (λ) and length(L).

$$\mathbf{K}_n = \frac{Mean\ Free\ Path}{Length} = \frac{\lambda}{L} \tag{2.10}$$

On scaling Knudsen number with 'L':

TABLE 2.1
Relationship of Reynolds Value with Nature of Fluid Flow

Reynolds(R_e) Value	Type of Flow	Observations
Re <2300	Laminar	Viscous force dominates
2300<Re>4000	Transition	Low flow rates
Re>4000	Turbulent	Inertial force dominates

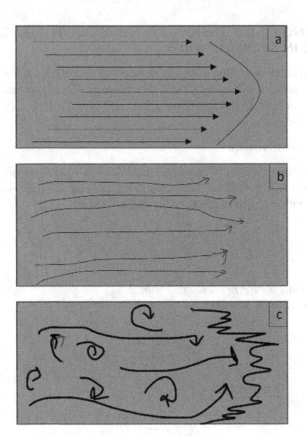

FIGURE 2.2 Illustration showing (a) laminar, (b) transition and (c) turbulent flow.

2.9.2.1 When velocity is constant:

$$K_n = \frac{\lambda}{L} \propto L^{-1} \tag{2.11}$$

2.9.2.2 When pressure difference is constant:

$$K_n = \frac{\lambda}{L} \propto L^{-1} \tag{2.12}$$

According to Zhu et.al, flow rates may vary on different types of substrates, so he reported a study of flow rates on different substrates and measured the contact angles and long life of capillary flow in PDMS and PC chips to give better meaning in measuring the flow rates [37]. The effects of channel aspect ratio and angles were studied for fluid flow [38]. Though the capillary and elements contribute many characteristics to microfluidics, they cannot deliver self-regulated flows for performing advanced functions [39]. The capillary action depends on surface tension of the analytes to the channel surfaces [40]. The fluid concentration in the outlet limits the overall flow duration; however, multiple outlets are used to reduce the concentration rate in each reservoir to maintain a longer flow duration [41].

2.10 DIFFERENT TYPES OF FLUIDS

Fluids can be classified into two types: Newtonian Fluids and Non-Newtonian Fluids.

- *Newtonian Fluids*: Fluids that obey Newtonian Law of Viscosity.
- *Non-Newtonian Fluids*: Fluids that do not obey Newtonian Law of Viscosity.

2.10.1 VISCOSITY

Viscosity is the friction exerted by the fluid to resist its flow.

2.10.2 ABSOLUTE VISCOSITY

It measures the internal resistance. The tangential force acting per unit area is required to move a horizontal plane w.r.t. another plane.

$$\mu = \tau \frac{dz}{du} \tag{2.13}$$

Therefore, from continuum mechanics, liquids and gases are fluids that obey equations of motion. For incompressible flows, the Reynolds Number is primarily a dimensionless parameter that determines the nature of the flow field. Still, if the Reynolds Number and geometry are matched, liquid and gas flows should be identical [42]. Since, for MEMS applications, we anticipate the possibility of non-equilibrium flow conditions and the consequential invalidity of the Navier–Stokes equations and the no-slip boundary conditions. Such circumstances can best be researched using the molecular approach.

2.11 NATURE-INSPIRED PHENOMENON: RECONFIGURATION OF PLANTS UNDER FLUID FLOW

Plants always rely on flexibility to change form and reduce drag when subjected to fluid flow. Instead of saying that plants deform, which carries a pathological meaning, Vogel (1984) mentioned the term Reconfiguration that describes the change of shape of plants under fluid flow. By bending and twisting, tree leaves, branches, and full crowns, as well as macro-phytes or seaweed blades, reduce their cross-sectional area normal to the flow and become more streamlined. At a high Reynolds number, a typical bluff body perceives a drag force proportional to the square of the flow velocity. Because plants reconfigure, their drag deviates from this scaling. We express this deviation with the Vogel exponent

$$F \propto U^{\infty 2+V} \tag{2.14}$$

where F is the drag force, $U\infty$ the free stream flow velocity away from disturbances, and V the Vogel exponent.

For example, the tulip tree leaf, reconfigures by rolling into an increasingly acute cone under increasing wind speed. This reduces its cross-section standard to the flow and makes it more streamlined. What comes across from the examples of the leaf and the tree crown is that three mechanisms are responsible for drag reduction by reconfiguration: area reduction, streamlining, and effective velocity reduction.

2.12   FABRICATION TECHNIQUES FOR MICROFLUIDICS DEVICES

The factors such as materials, design, and fabrication processes are selected to fabricate a micro-channel for a given application. Silicon was the first material to be worked upon for manufacturing microfluidic devices, followed by glass. However, materials such as polymers, composites, and paper have become well-known for fabricating micro-reactors. In translation from the laboratory to industrial scale, priority is given to production, performance, and reliability costs. Thus, a crucial aspect is to bridge the balance between the materials and fabrication techniques to bring down the product's final price. Established silicon wafer processing techniques, the resistance of silicon to organic solvents, and silicon's excellent physical properties have significantly contributed to the speedy growth of microfluidics technology. The methodology for fabricating silicon-based microfluidics devices generally begins with the substrate cleaning process. Techniques such as etching, lithography, electroplating, and molding (LIGA) are employed to manufacture microfluidic devices. Post-channel fabrication processes such as anodic and fusion bonding are of the utmost importance in closing the open channels. However, apart from being expensive, the issues of non-flexibility in silicon wafers, toxic chemicals used in the fabrication process and limited opacity are directly correlated to the downside of usage of silicon wafers for fabrication techniques. Thus, glass and polymers are suitable candidates to overcome conventional silicon-based microfluidics devices. Properties such as optical transparency, electrical insulation, chemical inertness, and the low cost of processing make them suitable candidates for device fabrication. Glass capillaries are also a promising option used in fabrication of capillary-based micro-reactors. Still, the assembly of such capillaries requires handling skills and cumbersome operations. A micro-channel on glass or quartz is fabricated via photolithography and wet or dry etching. Although glass is considered biocompatible and has widespread application in biological science, it is brittle and requires extensive care during the fabrication and handling of such devices. Similar to silicon-based microfluidics devices, the channels made in glass are also open channels and need a cleanroom environment at the time of bonding. Thus, during manufacturing, this adds to the overall cost of glass-based microfluidics devices. The underlying disadvantages of both silicon- and glass-based microfluidics devices have triggered extensive research in the field of finding alternate materials for device fabrication. The class of polymers used in the fabrication process is polydimethylsiloxane (PDMS), polystyrene (PS), polycarbonate (PC), and Polymethylmethacrylate (PMMA). Polymeric microfluidics devices gained popularity as they are low cost, optically transparent, biocompatible, disposable, and offer design flexibility. Fabrication techniques such as soft micromachining that includes laser ablation and micromachining, computer numerical control (CNC) micromachining, optical/photo/X-ray lithography, hot embossing, soft lithography, and injection molding is the technique employed for the fabrication of polymeric microfluidics devices. Recently 3D printing is also being used to fabricate COC-based microfluidics devices. Apart from inorganic and polymeric materials, paper is one such material that has become a promising option as a substrate for microfluidics devices. Besides being flexible, biocompatible, and most cost-effective, paper-based devices can be disposed of either by burning or in natural degradation. By the process of surface/ chemical modifications, capillary action on the surface of paper can be controlled.

Paper is an excellent option in microfluidics technology for biomedical analysis and forensic diagnostics. The detection or investigation of analytes using paper-based microfluidic devices is carried out through either colorimetric, electrochemical, chemi-luminescence, or electro-chemi-luminescence. Moreover, the paper's white background imparts excellent contrast for colorimetric detection. Due to paper's porous nature, paper-based devices have extended applications in the field of filtration and separation. Physical and chemical techniques, including inkjet printing, wax patterning, lithography, plasma treatment, and laser treatment, are used to fabricate paper-based Microfluidics devices. Paper origami and stacking are also techniques that are used to produce 3D paper-based microfluidic devices.

2.13 CONCLUSION

This chapter gives an insight into the fundamental laws and equations governing the fluid flow in a MEMS-based microfluidic device. These fluid flow equations are vital to an understanding before diving into any application related to this field. While scaling down the processes from macro-scale to micro-scale, it is essential to question and understand the fluid flow relationship because of the change in dimensional scale. Within micro-channels, the flow is usually laminar and thus hinders the mixing process. Specifically, when microfluidic devices are used, T-sensors which are used nowadays, play a crucial role in the diffusion mechanism. Therefore, the understanding of fluid flow has perhaps an edge in utilizing the concept of microfluidics for low-cost and rapid POC devices that can be operated without hiring trained professionals or setting up hefty laboratory infrastructure.

REFERENCES

1. C.R. Grayson, R.S. Shawgo, A.M. Johnson, N.T. Flynn, Yawen li, M.J. Cima, R. Langer, A BioMEMS review: MEMS technology for physiologically integrated devices, *Proc. IEEE*. 92(1):6–21, Jan 2004. DOI: 10.1109/JPROC.2003.820534.
2. M. Gad-el-Hak, Flow Control, ASME. *Appl. Mech. Rev.* 42(10):261–293, Oct 1989. https://doi.org/10.1115/1.3152376.
3. W.C. Tian, E. Finehout, Microfluidic diagnostic systems for the rapid detection and quantification of pathogens. *Microfluidics Biol. Appl.* 271–322, 2008. DOI: 10.1007/978-0-387-09480-9_9
4. R. Dong, Y. Liu, L. Mou, J. Deng, X. Jiang, Microfluidics-based biomaterials and biodevices. *Adv. Mat.* 31(45):1805033, Nov 2019.

5. C.M. Pandey, S. Augustine, S. Kumar, S. Kumar, S. Nara, S. Srivastava, B.D. Malhotra, Microfluidics based point-of-care diagnostics. *Biotechnol. J.* 13(1):1700047, Jan, 2018.
6. S.P. Reise, N.G. Waller, Item response theory and clinical measurement. *Ann. Rev. Clin. Psychol.* 5(1):27–48, Apr 27, 2009.
7. S.C. Terry, J.H. Jerman, J.B. Angell, A gas chromatographic air analyzer fabricated on a silicon wafer. *IEEE Trans. Electron Devices.* 26(12):1880–1886, Dec 1979.
8. H.A. Stone, S. Kim, Microfluidics: Basic issues, applications, and challenges. American Institute of Chemical Engineers. *AIChE J.* 47(6):1250, Jun 1, 2001.
9. M.H. Ansari, S. Hassan, A. Qurashi, F.A. Khanday, Microfluidic-integrated DNA nanobiosensors. *Biosens. Bioelectron.* 85:247–260, Nov 15, 2016.
10. A. Manz, N. Graber, H.Á.Widmer, Miniaturized total chemical analysis systems: A novel concept for chemical sensing. *Sens. Actuators B: Chem.* 1(1–6):244–248, Jan 1, 1990.
11. H. Bruus, *Theoretical microfluidics.* Oxford University Press, 2007.
12. C.D. Chin, V. Linder, S.K. Sia, Commercialization of microfluidic point-of-care diagnostic devices. *Lab on a Chip.* 12(12):2118–2134, 2012.
13. N. Convery, N. Gadegaard, 30 years of microfluidics. *Micro Nano Eng.* 2:76–91, Mar 1, 2019.
14. W. Jung, J. Han, J.W. Choi, C.H. Ahn, Point-of-care testing (POCT) diagnostic systems using microfluidic lab-on-a-chip technologies. *Microelectron. Eng.* 132:46–57, Jan 25, 2015.
15. P. Abgrall, A.M. Gue, Lab-on-chip technologies: Making a microfluidic network and coupling it into a complete microsystem—a review. *Journal of Micromech. and Microeng.* 17(5):R15, Apr 24, 2007.
16. W. Jung, J. Han, J.W. Choi, C.H. Ahn, Point-of-care testing (POCT) diagnostic systems using microfluidic lab-on-a-chip technologies. *Microelectron. Eng.* 132:46–57, Jan 25, 2015.
17. A. Heller, B. Feldman Electrochemical glucose sensors and their applications in diabetes management. *Chem. Rev.* 108(7):2482–2505, Jul 9, 2008.
18. L. Liu, C. Peng, Z. Jin, C. Xu, Development and evaluation of a rapid lateral flow immunochromatographic strip assay for screening 19-nortestosterone. *Biomed. Chromatogr.* 21(8):861–866, Aug 2007.
19. C. Tuerk, L. Gold, Systematic evolution of ligands by exponential enrichment: RNA ligands to bacteriophage T4 DNA polymerase. *Science.* 249(4968):505–510, Aug 3, 1990.
20. G.J. Kost, N.K. Tran, M. Tuntideelert, S. Kulrattanamaneeporn, N. Peungposop, Katrina, the tsunami, and point-of-care testing: Optimizing rapid response diagnosis in disasters. *Am. J. Clin. Pathol.* 126(4):513–520, Oct 1, 2006.
21. M.H. Wu, S.B. Huang, G.B. Lee, Microfluidic cell culture systems for drug research. *Lab on a Chip.*10(8):939–956, 2010.
22. L.Y. Wu, D. Di Carlo, L.P. Lee, Microfluidic self-assembly of tumor spheroids for anticancer drug discovery. *Biomed. Microdevices.* 10(2):197–202, Apr 2008.
23. L. Kang, B.G. Chung, R. Langer, A. Khademhosseini, Microfluidics for drug discovery and development: From target selection to product lifecycle management. *Drug Discov. Today.* 13(1–2):1–3, Jan 1, 2008.
24. C. Kleinstreuer, J. Li, J. Koo, Microfluidics of nano-drug delivery. *Int. J. Heat Mass Transf.* 51(23–24):5590–5597, Nov 1, 2008.
25. Z.B. Liu, Y. Zhang, J.J. Yu, A.F. Mak, Y. Li, M. Yang, A microfluidic chip with poly (ethylene glycol) hydrogel microarray on nanoporous alumina membrane for cell patterning and drug testing. *Sens. Actuators B: Chem.* 143(2):776–783, Jan 7, 2010.

26. Y.C. Toh, T.C. Lim, D. Tai, G. Xiao, D. van Noort, H. Yu, A microfluidic 3D hepato-cyte chip for drug toxicity testing. *Lab on a Chip.* 9(14):2026–2035, 2009.

27. L. Yu, M.C. Chen, K.C. Cheung, Droplet-based microfluidic system for multicellular tumor spheroid formation and anticancer drug testing. *Lab on a Chip.*10(18):2424–2432, 2010.

28. H.E. Abaci, K. Gledhill, Z. Guo, A.M. Christiano, M.L. Shuler, Pumpless microfluidic platform for drug testing on human skin equivalents. *Lab on a Chip.* 15(3):882–888, 2015.

29. S.R. Khetani, D.R. Berger, K.R. Ballinger, M.D. Davidson, C. Lin, B.R. Ware, Microengineered liver tissues for drug testing. *SLAS Technol.* 20(3):216–250, Jun 1, 2015.

30. J.M. Martel, M. Toner, Particle focusing in curved microfluidic channels. *Sci. Rep.* 3:3340, 2013.

31. J. Krüger, K. Singh, A. O'Neill, C. Jackson, A. Morrison, P. O'Brien, Development of a microfluidic device for fluorescence activated cell sorting. *J. Micromech. Microeng.* 12(4):486, Jun 19, 2002.

32. J.C. McDonald, D.C. Duffy, J.R. Anderson, D.T. Chiu, H. Wu, O.J. Schueller, G.M. Whitesides, Fabrication of microfluidic systems in poly (dimethylsiloxane). *Electrophoresis: Internat. J.* 21(1):27–40, Jan 1, 2000.

33. T.C. Merkel, V.I. Bondar, K. Nagai, B.D. Freeman, I. Pinnau, Gas sorption, diffusion, and permeation in poly (dimethylsiloxane). *J. Polym. Sci. Part B: Polym. Phy.* 38(3):415–434, Feb 1, 2000.

34. S. Yd, S.C. Maroo, Origin of surface-driven passive liquid flows. *Langmuir.* 32(34):8593–8597, Aug 30, 2016.

35. Y. Du, J. Shim, M. Vidula, M.J. Hancock, E. Lo, B.G. Chung, J.T.Borenstein, M. Khabiry, D.M. Cropek, A. Khademhosseini, Rapid generation of spatially and tempor-ally controllable long-range concentration gradients in a microfluidic device. *Lab on a Chip.* 9(6):761–767, 2009.

36. K. Iwai, R.D. Sochol, L.Lin, Finger-powered, pressure-driven microfluidic pump. *IEEE 24th International Conference on Micro Electro Mechanical Systems,* 1131–1134, Jan 23, 2011.

37. M.J. Davies, M.P. Marques, A.N. Radhakrishnan, Microfluidics theory in practice, in *Microfluidics in detection science,* 29–60, Royal Society of Chemistry, 2014.

38. S. Mukhopadhyay, S.S. Roy, A. Mathur, M. Tweedie, J.A. McLaughlin, Experimental study on capillary flow through polymer microchannel bends for microfluidic applications. *J. Micromech. Microeng.* 20(5):055018, Apr 14, 2010.

39. A. Olanrewaju, M. Beaugrand, M. Yafia, D. Juncker, Capillary microfluidics in microchannels: from microfluidic networks to capillaric circuits. *Lab on a Chip.* 18(16):2323–2347, 2018.

40. N.S. Lynn, D.S. Dandy, Passive microfluidic pumping using coupled capillary/evapor-ation effects. *Lab on a Chip.* 9(23):3422–3429, 2009.

41. D. Juncker, H. Schmid, U. Drechsler, H. Wolf, M. Wolf, B. Michel, N. de Rooij, E. Delamarche, Autonomous microfluidic capillary system. *Anal. Chem.* 74(24):6139–6144, Dec 15, 2002.

42. E.B. Arkilic, M.A. Schmidt, K.S. Breuer, Gaseous slip flow in long microchannels. *J. Microelectromech. Syst.* 6(2):167–178, Jun 1997.

3 Microneedles

New Tools of Painless Interface to Skin

3.1 INTRODUCTION

Drug delivery systems (DDS) are technologies that ensure efficient delivery and release of pharmacologically active compounds into cells, tissue, and organs. An ideal DDS should maximize therapeutic efficacy and minimize the side effects. There are various kinds of administration modalities, such as oral administration, transdermal administration, lung inhalations, mucosal administration, and intravenous injections. Among them, transdermal drug delivery systems (TDDS) are gaining popularity in biomedical applications.

The administration route for TDD is through the skin – the largest organ in the human body. TDD is the third-largest administration modality after oral administration and injections. The key advantage of TDD is the ease of drug transport through skin layers, which could reduce the fluctuations of drug levels in the blood, thereby minimizing the toxic side effects. Further, the given drug could bypass the first-pass metabolism of liver leading to good bioavailability of the drug to exert maximum therapeutic benefits. However, ironically, the major roadblock to TDD is the first layer of skin – stratum corneum (SC), which is a strong barrier for transport of molecules, drugs, and biological macromolecules transdermally. Therefore, it is pertinent to search for efficient TDD systems that may help to limit multiple application doses and improve patient compliance.

The microneedles in nature are difficult to notice, but often their effects remind us of their existence. Figure 3.1 shows such examples of microneedles in nature. Microneedles are important weapons of the bloodsucking bugs like mosquito, bed bugs, fleas and ticks. They exist in the protection system of porcupines as quills. Interestingly, a porcupine quill has a detachable barb that keeps its predators away. It is also seen in some species that the quills are easier to penetrate and difficult to remove. The Jumping Chola Cactus is covered with around 50 µm spines called glochids. They attach to slight brush by a passer-by. These spines are used by the plants to protect them against attacks. The farmers exercise caution while harvesting eggplants as the small spikes on the leaves and stem might hurt them. Whether it is for plant and animal protection or for survival, the microneedles have been successfully fulfilling the purpose for which they are present in each scenario. This is why the bio-inspired designs have time and again fuelled the imagination of engineers and inspired them. We shall take up one such example in the next section.

DOI: 10.1201/9781003202264-3

FIGURE 3.1 Examples of microneedle-like structures existing in nature. (a) Spines of North American Porcupine (*Erethizon dorsatum*) retrieved from https://nationalzoo.si.edu/news/two-north-american-porcupines-exhibit-national-zoo accessed on 24.04.21, (b) spines of Jumping Chola Cactus (*Cylinropuntia fulgida*) retrieved from https://succulentcity.com/are-cactus-thorns-poisonous/ accessed on 24.04.21, microneedle like weapons for blood sucking of (c) bed bug (*Cimicidae*) retrieved from www.esa.int/Applications/Observing_the_Earth/High_resolution_satellite_imagery_assists_hunt_for_infectious_kissing_bugs, (d) female mosquito (*Culicidae anopheles*). retrieved from www.terminix.com/blog/home-garden/blood-sucking-bugs/ and thorns on the leaves of eggplant retrieved from http://my.chicagobotanic.org/tag/plant/ accessed on 15.06.21.

3.2 BIO-INSPIRED DESIGN – FEMALE MOSQUITO

While designing a microscale system, often one is stuck on the choice of technology that actually works at this scale. It is a good idea to start with looking around us to get cues. Upon closely examining any of the examples discussed in the previous section, we see finesse of nature in building wonderful microneedle-based microfluidic systems. It will be relevant to take up the example of a female mosquito's blood-sucking mechanism here. The mosquito needle called the proboscis has a lumen which is only 25–30 μm in diameter (Figure 3.2c) and around 1500 μm length. Still, it operates well for transport of body fluids and blood. This diameter is quite small but large enough so as not to get clogged from the blood platelets (which are 20–25 μm in diameter). Interestingly, when a mosquito sucks blood, its labium is penetrated to the skin at the speed of 6–7 Hz by its head moving like a hammer. This vibratory motion allows for low insertion force (10–20 μN) for microneedles which also

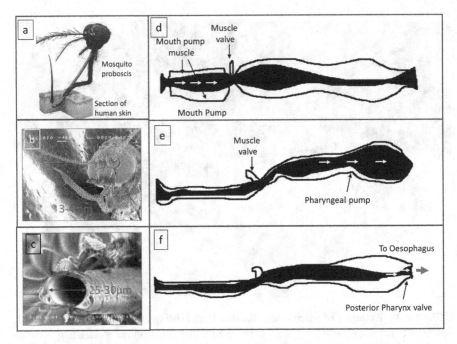

FIGURE 3.2 (a) Female mosquito proboscis, adapted from [3], (b–c) Scanning Electron Microscopy picture of female mosquito proboscis, [4] and (d–f) simple representation of sucking of blood by female mosquito adapted from [1].

account for less or no pain. The mosquito proboscis labum has an elasticity modulus of around 1.75 GPa and nanohardness of 0.08 GPa. The proboscis acts as inlet to a fully developed micropumping system lying in the mosquito's pharyngeal cavity. This three-step process is shown in a simplistic manner (Figure 3.2d–f). For the extraction method, first of all, a muscle valve is relaxed, a muscle of the mouth pump is tensed. Therefore, blood is extracted through the labium into the mouth pump. Secondly, the muscle of the mouth pump is relaxed, and the extracted blood is sent into the pharyngeal pump. Finally, the posterior pharynx valve is loosened. The extracted blood is sent into the esophagus. Thus, a female mosquito extracts blood at the amount of 1.9 µl for 2 minutes with pump operating with negative pressure [1,2].

Hence on studying a female mosquito's blood extraction mechanism, one can infer the following,

a. A device made up of microneedle, drug reservoir and micropump shall be suitable for diagnostic / therapeutic applications to human beings.

b. Microneedles with 35–40 µm diameter will not be clogged even if blood platelets enter the lumen.

c. Vibratory motion allows for low insertion force (10–20 µN) for microneedles which also account for less or no pain.

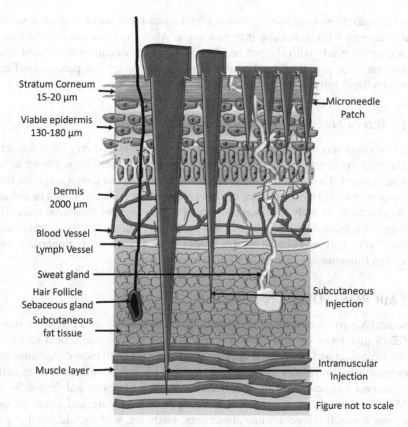

Stratum Corneum
15-20 µm

Viable epidermis
130-180 µm

Dermis
2000 µm

Blood Vessel

Lymph Vessel

Sweat gland

Hair Follicle

Sebaceous gland

Subcutaneous
fat tissue

Muscle layer

Microneedle
Patch

Subcutaneous
Injection

Intramuscular
Injection

Figure not to scale

FIGURE 3.3 Structure of skin and penetration of different kinds of needles in skin adapted from Maaden et.al. [7].

3.3 PAINLESS MICRONEEDLES

Microneedles can be defined as micron-sized structures which create tiny pathways in skin through which drugs can be delivered in a painless way. These microneedles penetrate the stratum corneum, the upper layer of the skin, and deliver the drug in the transdermal region of the skin. They barely touch the nerve endings present in this skin region and hence they cause little or no pain [5,6].

As the microneedles come into contact with skin, skin is stretched and results in high stress in skin around the microneedle. Skin starts bending downwards due to its elastic property. The insertion force keep on increasing till it reaches a maximum value. This happens due to the increase in resistance offered by skin. Then the skin tissue ruptures. Once the skin tissue is cut, insertion happens smoothly. The drug is delivered by the microneedles depending upon the microneedle type. Then the microneedle is retracted, and the skin tissues relax as microneedles are withdrawn. Chen et al. [8] studied the effect of different insertion angles of skin penetration and found that maximum insertion force was found when there was a vertical insertion and it decreased with increasing insertion angle. Also, the force required to pierce skin decreases greatly with sharper needle tips. If a microneedle array is used, then the insertion force is divided by the number of microneedles. The penetration force depends on the density of the microneedles.

3.3.1 BED OF NAILS EFFECT IN MICRONEEDLES

It's a common observation that when the microneedles are placed very close together, with interspacing less than 150 μm, they fail to pierce the skin. This is known as the bed of nails effect. This phenomenon got its name after a famous Indian magician trick where a person would sleep on a bed of closely spaced big nails and would be unhurt. This was because his body load was distributed in the very closely spaced nails. The resulting force exerted by a single needle would then be less than the required skin piercing force. This was shown by Lhernould et al. [9] where microneedles with larger pitch (interspacing) were more effective in skin insertion.

3.4 MICRONEEDLES ON TIMELINE

Biomedical Micro-Electro-Mechanical Systems (Bio-MEMS) are a particular subset of MEMS that focus on fabrication of miniaturized electro-mechanical technologies for biological and biomedical applications – which due to their dimensions and biocompatibility can rapidly integrate with various biological entities such as cells, tissues, nerves and veins, making the system suitable for transdermal drug delivery. Bio-MEMS technologies are highly interdisciplinary in nature and combine vast areas from material sciences, clinical sciences, medicine, surgery, electrical engineering, mechanical engineering, optical engineering, chemical engineering, and biomedical engineering. Some of the major applications of Bio-MEMS include genomics, proteomics, molecular diagnostics, point-of-care (POC) diagnostics, tissue engineering, single cell analysis and implantable microscopic biomedical sensors. Examples of Bio-MEMS devices that are already available include DNA microarray chips for multiple analyses of DNA and microneedle arrays (MNAs) for non-invasive and painless drug delivery and detection of a variety of specific analytes simultaneously in samples such as blood, urine and sputum.

Microneedle array-based Bio-MEMS are primarily applied for active transdermal drug delivery and have the potential to emerge as an effective, minimally invasive and thereby painless replacement to traditional hypodermic syringe-based injections. Microneedle arrays of height ranging between 25 to 2000 μm penetrate the stratum corneum of the skin and deliver the drug with a minimally invasive and painless action – without contacting any pain receptors present in the dermis of the skin at depths of 1.5 to 4 mm. Microneedles have been used for various applications including drug and vaccine delivery, and disease diagnostics.

The microneedle research has come a long way since its first patent in 1976. The microneedle evolution can be categorized into three generations. They are

1. **1st generation microneedle** – In the first generation of microneedle research (1976–2010), numerous fabrication techniques and materials were used for fabrication. Here, emphasis was to show the microneedle as a novel and efficient tool for skin penetration. It also covered the preliminary findings on delivery of key drugs through microneedles.

2. **2nd generation microneedle** – The second generation of microneedles (2011–2016) are categorized by exploration of a batch manufactureable process for microneedle fabrication. Different components like micropump and different kinds of applicators were integrated with microneedle patches. Microneedles started getting researched for a variety of applications. Most of the patents granted were mostly related to microneedle fabrication. Microneedle patches were commercially launched mostly in the cosmetic sector.

3. **3rd generation microneedle** – The third generation of microneedles (2016 onwards) is the current stage for the microneedles where they developed into a smart device integrated with different components like actuators, micropump or applicators. Their fabrication began to be preferred mainly by processes which can mass manufacture them. Apart from the therapeutic field, the application area of microneedles expanded to a cosmetic, prophylactic and diagnostic tool. The focus of microneedle patents shifted to application areas like vaccine delivery.

3.5 TYPES OF MICRONEEDLES

The key to success for drug delivery through microneedles is fourfold. They are: development of high strength MNs to pierce the skin, optimum drug flow rate, minimization of side effects and no introduction of toxicity [10]. Microneedles can be of different shapes and structures for the main function of creating a pathway in skin for therapeutics delivery. The microneedles may be in plane or out of plane or present in an array to assist the rapid drug delivery across the skin [11]. Based upon functional structure, microneedles can be mainly divided into two categories (Figure 3.4).

(i) Solid microneedles, and
(ii) Hollow microneedles

3.5.1 SOLID MICRONEEDLES

These microneedles are solid in structure and apply drugs in a variety of ways. The advantages of solid microneedles are that they are easier to fabricate since they do not require elaborate fabrication processes. Their disadvantage is that they can deliver limited drug load only. Over the years, we have seen different kinds of solid microneedles. They can be categorized as

(i) Poke and patch microneedles
(ii) Coated microneedles
(iii) Dissolvable microneedles

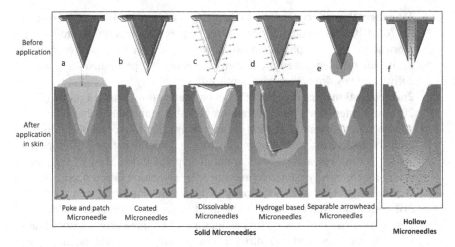

FIGURE 3.4 Illustration showing the functionality of different types of microneedles. Microneedles are mainly categorized into solid and hollow microneedles depending upon their structure. Solid microneedles can be further categorized into (a) poke and patch microneedle, (b) coated microneedles, (c) dissolvable microneedles, (d) Hydrogel based microneedles and (e) Separable arrowhead microneedles where a limited quantity of drug can be administered by various means. On the other side hollow microneedles (f) have a hollow lumen through which can deliver relatively larger amount of fluids.

(iv) Hydrogel-based microneedles
(v) Seperable arrowhead microneedles
(vi) Hollow microneedles

3.5.1.1 "Poke and Patch" Microneedles

In this kind of microneedles, micro-sized incisions are created in skin using the microneedles and they are removed (Figure 3.4a). Pointed tips of the microneedles penetrate the skin and create channels of micrometer size, through which the drug directly enters the skin layers. A drug patch is then applied to the skin. The intratransdermal pathway created by the microneedles aids absorption of drugs in skin. They deliver a limited drug load and could be used for sustained release. Solid microneedles deliver the drug with passive diffusion to skin layers. This type of microneedle structure is designed to penetrate the stratum corneum in order to enhance drug delivery to the dermis to improve the bioavailability and kinetic transport across the skin. In comparison to intramuscular delivery, the solid microneedle is suitable for delivery of vaccines as it lasts longer and possesses a more robust antibody response.

3.5.1.2 Coated Microneedles

In another variant of microneedles, drugs could be pre-coated on the solid microneedles (Figure 3.4b). Such microneedles are called coated microneedles. These microneedles are surrounded with the drug solution or drug dispersion layer.

The amount of drug that can be loaded depends on the thickness of the coating layer and the size of the needle that is penetrating the skin. A coated microneedle can deliver proteins and DNA into the skin with a minimally invasive manner. A patch containing microneedle array could be inserted in skin where the drug diffuses into the body with time. After some time, this patch could be removed. They deliver drugs immediately and require only short application time. The drug can be coated by a variety of processes like dip coating, spraying etc. An advantage of a coated microneedle is rapid delivery of the drug to the skin; however, the reusability of such needles is questionable due to their tendency to infect a new set of patients due to the presence of residual remains of the drug administered to previous patients.

3.5.1.3   Dissolvable Microneedles

Another variant of solid microneedles which has been extremely popular over the years is dissolvable microneedles (Figure 3.4c). They are made using soluble drug matrices [12].Various polymers like poly(vinyl alcohol)(PVA), dextran, carboxymethyl cellulose(CMC), etc. have been used for fabrication of such MNs. Dissolving microneedles are fabricated with biodegradable polymers by encapsulating the drug into the polymer. The dissolving microneedles are a combination of needle structural material and solid drug formulation. This kind of patch comes with a backing material to the microneedle array which could be immediately removed after application. The structural material allows skin penetration of the drug and upon insertion the drug comes into contact with body fluids. This leads to softening of the total microneedle and it starts getting dissolved in body fluids. From there it is picked up by the veins and systemic circulation of the body. The application involves only a single step as the microneedle is not to be removed after insertion as in other cases. The polymer gets de-graded inside the skin and controls the drug release. The bio-acceptability and dissolution of the polymer inside the skin make it one of the best choices for long-term therapy with improved patient compliance. Due to improvement observed in applying dissolvable microneedles following "poke-and-release", this approach is considered better than other approaches. The dissolvable microneedle tip can be loaded in a timely manner via a two-step casting method. Upon insertion of the dissolvable microneedle into the skin, the drug load releases and diffuses easily by dissolution of the needle tip. Water-soluble materials are most appropriate for the manufacture of the dissolvable microneedle. Likewise, the micro-mold method of fabrication is most suitable to produce dissolvable microneedles. The main disadvantage of this method is that it deposits undesirable polymers or substances in body [13].

3.5.1.4   Hydrogel Forming Microneedles

Hydrogel forming MNs (Figure 3.4c) are a relatively new class of MNs where the needles swell when they come into contact with body interstitial fluids after insertion and aid in diffusion of the drug through them (Figure 3.4d). Their advantage is that they can carry a higher drug load than the other types of microneedles and for a sustained period. Such MNs have been preferred, but their disadvantages are that they soften on swelling and may break in skin during needle retraction and are limited by diffusion in skin [14].

3.5.1.5 Separable Arrowhead Microneedles

This kind of microneedles are an adaptation of dissolving microneedles. The microneedle consists of two parts, a solid microneedle and a sharp dissolvable material tip on top of the solid microneedle (Figure 3.4e). This kind of arrangement utilizes the strength of the metal microneedles as a base and takes a dissolvable tip to the desired depth in skin where the drug dissolution happens.

3.5.2 Hollow Microneedles

As the name suggests, hollow microneedles have a lumen or hollow pathway through their structure (Figure 3.4f). For these kinds of microneedles, liquid drug formulations could be stored in a reservoir above the microneedle array and the required amount of drug could be delivered. The advantages of hollow microneedles are that they enable delivery of a large amount of drugs compared to solid microneedles. Hollow MNs offer the advantages of offering large volume or continuous flow of drugs with control over flowrate. These kinds of microneedles are also preferred in devices where a microneedle is to be used both for a diagnostic as well as therapeutic purpose. The drug flow through the microneedle may be pressurized by a pump, can be simple diffusion-based or electrically actuating a membrane to push the drug [15]. Silicon, glass ceramics and polymers such as SU-8 and PLGA have been used in hollow MNs fabrication [16]. The disadvantage of using hollow MNs is that they require an elaborate fabrication process and only liquid drug formulations can be used [17].

After having known about the type of microneedles, we shall have a broad look at the materials which have been used to fabricate the microneedles.

3.6 STRUCTURAL MATERIALS FOR MICRONEEDLE FABRICATION

The selection of materials become important for microneedles as the material should have sufficient strength for skin penetration, it should be relatively inert so as not to react with drug molecules and should allow for immediate effective release of drugs. The choice of structural material for the microneedle also defines the shelf life and the applicable drug formulations [18]. There are several other parameters which need to be considered while choosing the MN material. When an MN is inserted into skin, then due to mishandling and an uneven skin surface, there are chances that the needle tip might break in skin. This tip material remaining in skin should not become a source of bad immunogenic effects. Hence the material chosen for MN material should be biocompatible [19]. It's also important to understand the role of substrate material for microneedles here. The interfacing of the microneedle array with the substrate should be such that there is good adhesion because we do not want the needles to break in skin upon retraction. Also, it should handle the mechanical stresses introduced due to improper handling of the patch and stresses arising due to day-to-day handling and storage of the patches. Considering these factors at the design stage decreases the commercialization time for the product [20]. Hence, the microneedles should be on such a substrate which can handle these stresses and is also compatible with the microfabrication processes. For most of the microneedles, the substrate is the same as the microneedle material, but in some cases it might be a different material altogether, like glass, stainless steel, silicon or some flexible material. Silicon is ideally suited as a substrate which is sturdy and can be micromachined [21]. One must essentially consider the thermal coefficients mismatch between the microneedle material and substrate since a large thermal mismatch may result in peeling of microneedles from substrate during fabrication.

During the last few decades, microneedle research has followed a trend. The trend for microneedle research in terms of its type and materials for the last 20 years is shown as Figure 3.5. The trend shows that fabrication of solid microneedles has been preferred as it is simpler than hollow microneedles. In the solid microneedles category, various types of microneedles emerged as we discussed above, thus expanding the research in this particular field. Research into hollow microneedles, though less, did not lose its charm over the years because of the potential of hollow microneedles to completely take over syringe needles. Silicon has its special place in both solid and hollow microneedles because it is a favourite of the MEMS industry. The materials used for microneedles are discussed below.

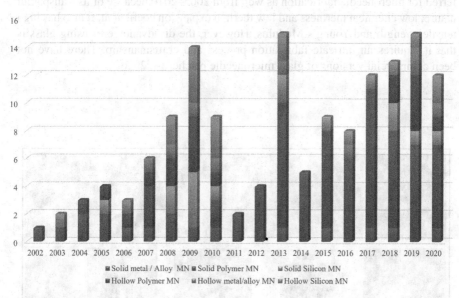

FIGURE 3.5 Chart showing microneedle research trend for last 20 years. It indicates the preference towards solid polymer microneedles where a variety of applications can be addressed and a consistent effort towards intricate fabrication of silicon and polymer hollow microneedles. Publication data courtesy google search engine with keywords solid microneedles, silicon microneedles, microneedles, and review articles Cárcamo-Martínez et al. [22], Yang et al. [23] and Moreira et al. [24].

Some of the established materials preferred for microneedle fabrication are listed below.

3.6.1 SILICON

Silicon has been most widely used in hollow MN fabrication because it remains the preferred MEMS material in the microfabrication industry [19]). Shikida et al. [25] used a quite elaborate process and developed MNs with flow channels by combination of mechanical dicing and anisotropic wet etching. Table 3 briefly covers the recent work done on silicon MNs. Chen et al. [26] fabricated solid silicon MNs by a reactive ion process and then formed biodegradable tips of porous silicon only on the tips of MNs. Coulman et al. [27] showed permeation of polystyrene nanoparticles in human skin by silicon microneedles for enhancing the transport of vaccine molecules across skin. In recent years, deep reactive ion etching (DRIE) emerged among other techniques for generation of complex shapes of silicon microneedles as shown by works of Liu et al. [28] and Vinaykumar et al. [29]. The biocompatibility of silicon is still debatable [30,31].

3.6.2 GLASS

Glass has been long used and accepted in the medical industry. It was initially preferred for microneedle fabrication as well from 2005–2010 because of its transparent nature, low chemical inertness and low thermal expansion coefficient. It also has high tensile strength and Young's Modulus. However, the disadvantage of using glass is that it requires an intricate fabrication process and craftsmanship. There have not been commercial versions of glass microneedle patches till 2020.

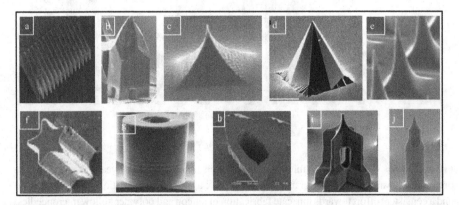

FIGURE 3.6 Scanning electron microscopy images of different geometries of solid and hollow silicon microneedles, (a) [32], (b) [25], (c) [26], (d) [27], (e) [33], (f) [34], (g) [28], (h) [35], (i) [16] and (j) courtesy of Debiotech SA, Lausanne, Switzerland are shown.

3.6.3 METALS AND ALLOYS

Metals and alloys have been preferred in making microneedles due to their high Young's Modulus, tensile strength, ductility and malleability. Nickel, iron, cobalt, palladium and titanium are some of the widely used metals for microneedle fabrication [36]. Stainless steel is another material which has been widely accepted as biocompatible material for medical devices. Solid and hollow microneedles have been fabricated from stainless steel but the micromachining of steel still remains a challenge. Apart from silicon and steel, metal MNs were also explored where once a silicon or polymer mold was formed, a process of electroplating can be adopted to make metallic MNs.

3.6.4 CERAMICS

Ceramics are a class of materials which possess high mechanical strength. Some of them are biocompatible too. Their disadvantage is that they are brittle. The biocompatible ceramics used for microneedle fabrication are Alumina and Zirconia.

3.6.5 POLYMERS

The polymer microneedles are preferred over other microneedles as they provide an inexpensive, biocompatible material that can easily be integrated into mass production

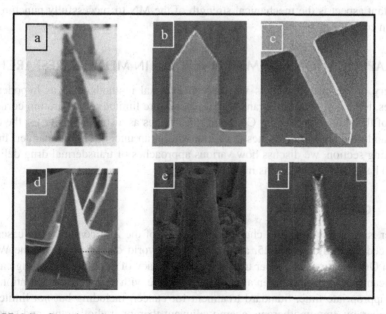

FIGURE 3.7 Scanning electron microscopy images of different geometries of solid and hollow metal / stainless steel microneedles (a) [37], (b) [38], (c) [15], (d) [29], (e) [30] and (f) [40].

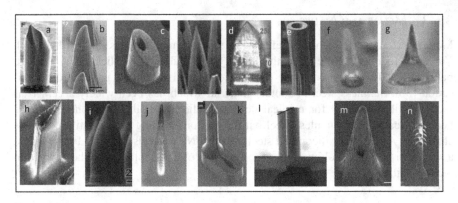

FIGURE 3.8 Hollow and solid polymer microneedles fabricated from polymer materials over the years, (a) [41], (b) [42], (c) [43], (d) [44], (e) [45], (f) [46], (g) [14], (h) [47], (i) [48], (j) [44], (k) [49], (l) [50], (m) [51], and (n) [52].

[53]. The use of polymers in MN fabrication has also attracted great attention due to their high mechanical strength and thermal resistance. With these advantages, organic (polymers which contain carbon in their chain) as well as inorganic (polymers which do not contain carbon in their chain) polymers like polyvinylpyrrolidone (PVP), polydimethylsiloxane (PDMS), parylene, polytetrafluoroethylene (PTFE), poly (methymetahcrylate) (PMMA), carboxymethyl cellulose (CMC), polylactide-co-glycolide (PLGA) and SU-8 have been widely used in MN fabrication [19]. The next important aspect is the mechanical strength of the MN to successfully puncture the stratum corneum.

3.7 APPLICATIONS OF MICRONEEDLES IN MEDICAL RESEARCH

Delivery of therapeutic agents via the conventional methods such as hypodermic needles, topical creams, and transdermal patches are limited to the stratum corneum layer of the skin. The Stratum Corneum (SC) serves as a strong barrier for the drug entry and thus less drug reaches the action site in an uncontrollable manner. In the following section, we discuss how various approaches of transdermal drug delivery (TDD) have helped physicians in disease therapy.

3.7.1 CANCER THERAPY

Cancer is the most clinically challenging disease of the 21st century, with cases set to exceed 20 million by 2025, according to the world cancer report by the World Health Organization. In cancer treatment, the efficacy of an anti-cancer drug can be significantly improved if given at the cancerous site with a proper concentration and at an appropriate time. Standard treatment for cancer, including but not limited to chemotherapy, immunotherapy, chemo adjuvant therapy, radiotherapy, and surgery, are known to be associated with severe side effects and low efficacy. The introduction of microneedles in cancer therapy as a synergistic combination of a transdermal drug

TABLE 3.1

Comparison of Polymers Commonly Used in Medical Devices and Microneedle Fabrication

Polymer	Abbreviation / Reference	Young's Modulus (GPa)	Tensile Strength (MPa)	Biocompatibility	Biodegradibility	Preferred Fabrication Method	Reference
SU-8	SU-8	4	34	Y	N	UV photolithography / Direct laser writing	[59]
High density polyethylene	HDPE	1.4	32	Y	N	Soft Moulding	[60], retrieved from www.accudynetest.com [65]
Poly(vinyl pyrrolidone)	PVP	1.6	20.8	Y	N	Moulding	[61]
Poly vinylidene fluoride	PVDF	2.1	50	Y	N	Soft Moulding	[62,63]
Poly(vinyl alcohol)	PVA	2.3	64.8	Y	Y	Soft Moulding	[64]
Poly(dimethyl siloxane)	PDMS	2.8	2.24	Y	N	Soft Moulding	[65]
Poly(methyl-methacrylate)	PMMA	3.1	76	Y	N	Soft Moulding	[65]
Chloro-poly-para-xylene	Parylene C	3.2	45	Y	N	Soft Moulding	[66]
Poly-lactic-co-glycolic acid	PLGA	7	30	Y	Y	Soft Moulding	[65]

(Continued)

TABLE 3.1 (Continued)
Comparison of Polymers Commonly Used in Medical Devices and Microneedle Fabrication

Polymer	Abbreviation / Reference	Young's Modulus (GPa)	Tensile Strength (MPa)	Biocompatibility	Biodegradibility	Preferred Fabrication Method	Reference
Poly-lactic-acid	L-PLA	0.1	66	Y	Y	Soft Moulding	[67], retrieved from www.creativemechanisms.com/blog/learn-about-polylactic-acid-pla-prototypes
Gelatin methacryloyl	GelMA	0.001	0.5	Y	Y	UV irradiation	[67], [68]
Sodium Hyaluronate	HA	0.002	0.06	Y	Y	Moulding / casting	[67]
Polyetheretherketone	PEEK	3.9	70	Y	N	Injection Moulding	[69]
Cyclic Olefin Copolymer	COC	3.2	46	Y	N	Injection Moulding	[70]

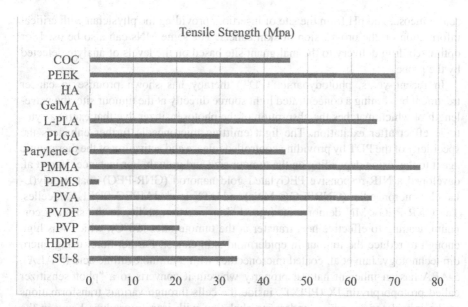

FIGURE 3.9 Chart showing comparing tensile strength of polymers which are used for microneedle fabrication or are potential candidates.

patch along with hypodermic needles, wherein an anti-cancer drug can be delivered directly to the tumour site by the microchannels of microneedles can significantly reduce the side effects of existing therapies and increase the prognosis substantially. Although microneedle-based drug delivery may not help all types of cancer it can considerably improve the treatment strategies for skin, breast, prostate, and cervical cancers with open drug administration sites.

In recent years, microneedles with sensor technology have been developed with an analyte recognition element and a transducer (electrochemical or optical). This helps in correctly optimizing the drug concentration at the site of delivery and can also detect the levels of essential biomolecules in the interstitial fluid. Hahn and co-workers in 2015 developed and patented a prototype of microneedles with electrochemical sensors that could detect the changes in nitrogen monoxide, which is a crucial product of the innate immune response of the body. The device was able to detect tumour size and growth. The microneedles are fabricated using a combination of polymeric bases made of adhesive polymers (chitosan, fibronectin, vitronectin, polydopamine, silk, and collagen) and conductive polymers (polyacetylene, polypyrrole, among others). Porphyrin or heme as iron ions is added as a layer to detect nitrogen monoxide specifically. These patented microneedle patches have the potential to be implemented in the therapy of breast, lung, uterine, and colorectal carcinoma. Another example of this "sense-act-treat" technology is a microneedle array developed by a joint cooperation between the "Regents of the University of California", North Carolina State University, and Sandia Corporation. A specific biochemical sensor or probe was attached to the needles, which has the ability to detect biomolecules such as lactic acid, glucose, and pH from the site of insertion, providing the physician with critical information on the progression of skin cancer. The same MNs can also be used for optimized drug delivery to the malignant site based on the levels of analyte detected by the probe.

In recent years, photodynamic (PDT) therapy has shown promise in cancer treatment by shining a concentrated light source directly at the tumour site, the wavelength of which matches the absorption of the photosensitizer dye that exerts a cytotoxic effect after excitation. The light emitting microneedle further enhances the specificity of the PDT by providing controlled release and activation of the therapeutic agent to the lesion depending on the tumour size and growth. For instance, Hao et al. developed a NIR-responsive PEGylated gold nanorod (GNR-PEG)-coated poly(L-lactide) microneedles (GNRPEG@ MNs) and DTX-loaded MPEG-PDLLA micelles. The GNR-PEG@MN demonstrated good skin insertion ability to the stratum corneum, leading to effective heat transfer at the tumour sites (50°C), which was high enough to reduce the tumour in epidermoid skin cancer. Furthermore, using micro dip technology, Jain et al. coated microneedles with 5-aminolevulinic acid (5-ALA). 5-ALA has an inherent natural property wherein it converts to a "photosensitizer called protoporphyrin IX (PPIX)", inside the cells through various transformations in the mitochondria, which undergoes chelation with iron to produce heme in the presence of an enzyme ferro-chelatase. The coated microneedle patches significantly suppressed tumor growth in porcine skin malignancy compared to the group that received tropical administration of 5-ALA.

Alongside chemotherapeutic approaches, immunotherapies are known to have better outcomes for skin cancer patients. Sustained delivery of immune checkpoint inhibitors such as anti-programmed death-1 (aPD1) antibody was achieved by Zheng et al. [54] by glucose oxidase (Gox) loaded pH-sensitive nanoparticles (NPs) as the Gox converted blood glucose to gluconic acid. This system was added successfully to hyaluronic acid-based dissolvable MNs to perform melanoma therapy and had remarkable anti-tumor efficacy and enhanced retention of aPD1 in the tumour sites.

It can be inferred from these examples and ongoing studies that different cancer-indicating biomarkers can be detected using microneedle-based cancer detection. Also, anti-cancer drugs can be targeted at the tumour sites efficiently based on microneedles.

3.7.2 Painless Vaccines

The dissolvable MNs were used to replace hypodermic injection needles that were typically used to administer vaccines. Unlike other types of microneedle, the dissolvable microneedles are biocompatible, robust, scalable, and do not generate biohazardous waste. Dissolvable microneedles were used to deliver vaccines for malaria, diphtheria, influenza, Hepatitis B, HIV, and polio.

Even though dissolvable microneedles are most frequently used for vaccine delivery, coated MNs arrays have also been successfully used for vaccination purposes. A study used a simple, safe, and compliant vaccination method to improve the immune system for pigs by administering bacillus Calmette–Guérin (BCG) vaccine with a coated microneedle. Another study successfully encoded hepatitis C virus protein in DNA vaccine coated on microneedle. The microneedle was effectively primed for specific cytotoxic T lymphocytes (CTLs) in mice. Furthermore, a coated microneedle carried influenza virus antigen for vaccination application in mice.

Hollow microneedles have been used to deliver anthrax recombinant protective antigen vaccine to a rabbit instead of regular injection. A hollow microneedle was evaluated for vaccination against plaque in a mouse model. A clinical trial conducted in humans using hollow microneedle with influenza vaccination showed similar results with the immune system when compared to intramuscular injection [55].

3.7.3 Diabetes Treatment

Diabetes Mellitus is a serious health condition affecting 422 million people world-wide, and 1.5 million deaths are attributed to diabetes every year. Subcutaneous insulin shot is the standard care to manage diabetes. However, insulin injections do not accurately release insulin into the body leading to inadequate glycaemic control and subsequent negative consequences such as blindness and multi-organ failure. Also, an overdose causes hypoglycaemia resulting in seizures, loss of consciousness, and even death. Hence, the development of MN-based insulin delivery approaches is underway to find an efficient system of drug delivery.

Dissolving microneedle patches made up of starch and gelatin was used to release insulin in diabetic rats [56]. The insulin-loaded microneedle patches applied could maintain stability at 25–37°C for at least a month and produced significant hypoglycaemic effects in diabetic rats. An improvement on the fabrication was done by Liu et al. [46], where a HA tip loaded microneedle array was used for TDD of exedin-4 in type-2 diabetic GK/Slc rats. The TDD was successful without any skin damage. Further, Yu et al. [57] were able to significantly reduce the pain in diabetic rats by fabricating a polymer-based microneedle (alginate and hyaluronate), which showed excellent mechanical strength and could release insulin in deeper tissues. In another study, Chen et al. [58], tried to overcome the incomplete insertion of the microneedles at the site of placement by designing microneedles with supporting structures of PVA/PVP and y-PGA microneedle tips. This provided the added advantage of extending skin length insertion importance for low dosage drugs. Taken together, it can be inferred that microneedle delivery systems have achieved some significant results which paved the way for human clinical trials and further studies on controlled release of insulin using microneedle arrays.

An important aspect of insulin delivery is to achieve control release of insulin an optimum concentration for personalized diabetes management. To address this problem, Yu et al. designed a novel microneedle patch for glucose responsive insulin delivery. Glucose oxidase enzyme (Gox) and insulin loaded glucose responsive vesicles (GRVs) were introduced as a component in the fabrication process. The key constituents of GRVs were hypoxia sensitive hyaluronic acid (HS-HA) conjugated with 2-nitroimidazole (NI). During hypoxic condition, NI could be converted into hydrophilic 2-aminoimidazoles via reduction reactions. This leads to dissociation of GRVs and release of insulin. In other words, the GRV element could release insulin in response to the changes in the blood glucose concentration, thereby regulating glucose levels in the blood. However, Gox produces a lot of H_2O_2 by converting glucose into gluconic acid, which may cause inflammation and rashes in the skin.

To reduce the undesirable effects of hydrogen peroxide Yu et al. designed hypoxia and H_2O_2 dual sensitive vesicles to deliver insulin. The dual sensitive vesicle element was composed of PEG and Polyserine modified with 2-NI via a thioether moiety (PEG-ply(Ser-S-NI), which contains an H_2O_2-sensitive moiety thioether. The diblock copolymer could be self-assembled into a nanoparticle to load Gox and insulin. Under hypoxic condition, the hydrophobic 2-NI–PEG-poly(Ser-S-NI) could be converted into hydrophilic 2-aminoimidazole leading to insulin release upon vehicle dissociation.

3.8 MICRONEEDLE APPLICATORS

The microneedle arrays are often housed in an arrangement which helps them to be applied to skin or gives them connection to the fluidic reservoir in case of hollow microneedles. The microneedle assembly helps to press the needles uniformly and effectively on skin. They also act as a protective housing for the microneedles. The assembly also aids in retracting the microneedle from skin without breaking. Microneedles are seen to be housed in different kinds of assemblies. They can be mainly categorized as follows.

3.7.1 ADHESIVE PATCH

Most common designs for applying coated, dissolvable and hydrogel-based microneedles is in the form of patches. The microneedle array is mounted on a skin adhesive patch and could be applied like adhesive bandage available widely. The patches are applied and then a slight thumb pressure pushes the microneedles into skin. They are left there for a recommended time and then removed by peeling (Figure 3.10a).

3.7.2 SYRINGE

Many times, the microneedles are available to be attached through an adaptor with a syringe. This arrangement is particularly required for hollow microneedles. The syringe acts as a reservoir and aids in delivery of a measured quantity of drug. The drug could be delivered by pushing the plunger, as is done for a conventional syringe. A modified syringe structure is seen in microneedle assembly from Debiotech (Figure 3.10b). The microneedles are attached in a polymer casing which has fluidic connection from adjacent syringe like barrel. In this part the plunger could be pushed to deliver the drug through the microneedle. The user is able to push a known volume of drug into skin just like the syringes.

3.7.3 PATCH APPLICATOR

The microneedle array could be housed in an applicator which aids in its application in skin. The applicator may leave the patch on skin and retract or it may be required to

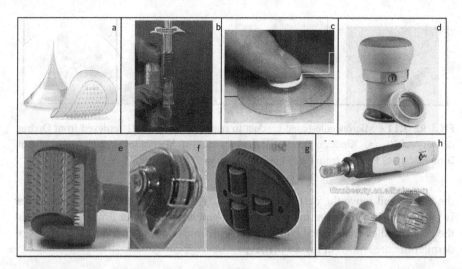

FIGURE 3.10 An overview of different kinds of microneedle assembly (a) Acropass from Raphas, retrieved from www.raphas.com/en/tech/core#1, (b) Microneedle assembly, DebioJect™. Courtesy of Debiotech SA, Lausanne, Switzerland, (c) MicroCor™Microneedle device MarquezGrana et al. [72] (d) Zosano Pharma's intracuteneous microneedle system retrieved from Ondrug delivery, (e-g) Different dermaroller designs with microneedles, Yang et al. [73] and (h) Microneedle pen retrieved from www.titanbeauty.en.alibaba.com

be pressed throughout the injection process. The microneedle array from Microcor™ is fixed inwards in the centre of an elastic concave shaped membrane as shown in Figure 3.10c. On the patch application, first the skin contact zone of the elastic membrane comes into contact with skin and a slight vacuum zone is created. Then the microneedle area is pressed to insert the microneedles vertically in skin thereby delivering the drug. The large elastic surface area of the membrane makes it easy for the patch to be pulled from skin. Zosano Pharma's intracuteneous microneedle system consists of an applicator device which houses the microneedle. The outermost part of the device has a protective covering. The microneedles are housed on a releasable microprotrusion membrane. The device can be pressed to insert the microneedle patch onto skin and then retract simultaneously leaving it there. This design minimizes the manual intervention in inserting the patch. After use, the patch could be removed manually.

3.7.4 Microneedle Roller

A series of innovations can be seen in the microneedle applicator called a dermaroller. They are essentially used for cosmetic purposes where solid microneedles pierce the skin creating microincisions or trauma to skin and the skin's healing process becomes active, thus regenerating the skin. The microneedles in a dermaroller are mounted on a cylinder and the cylinder is free to roll on the axle of the handle. One can easily hold

FIGURE 3.11 Microneedle-based product to reach market in 2021 retrieved from Ondrug delivery 2020, (a) Kindeva Drug Delivery device, (b) Microneedle patch from Micron Biomedical, (c) Product from Latch medical.

the handle and move the dermaroller on skin. The handle of the microneedle roller is pressed against the skin, inducing the movement of the roller drum and such that the microneedles are inserted into the skin with uniform pressure. Variants designs of dermaroller are shown in Figure 3.10e–g.

3.7.5 MICRONEEDLE PEN AND HANDLES

Another design of microneedle assembly is seen in form of pens and handles which have the drug reservoir and the microneedle patch integrated in them (Figure 3.10h). A common pen-like microneedle assembly is seen in the case of Dermapen and handle-shape assembly is seen in the case of the Kindeva hollow transdermal system (h-MTS). Kindeva h-MTS allows the user to insert the microneedles anywhere on the skin. The button on the device controls the force applied to the skin during injection.

3.8 COMMERCIALIZATION OF MICRONEEDLES

The proprietary dissolvable microneedle patch from Micron Biomedical Inc., US has micron-scale dissolvable needles which offer sustained release of water-soluble drugs in skin. It is applied using the thumb and has an indicator that provides the indication that the patch has been inserted successfully. This patch contains a thermostable drug formulation which has good thermostability for many biotherapeutics thus eliminating the need for a cold chain. There is no sharp waste as the microneedles dissolve in skin.

Another company, Raphas Co. Ltd., from South Korea, has several patents related to dissolvable microneedles, mainly in the cosmetics segment. It uses a very interesting technology of droplet-born air blowing technology for microneedle fabrication. Acropass and Theropass are two of the many commercial patches launched by them.

Nemaura Pharma Ltd., UK, has targeted osteoporosis therapeutics by using teriparadite coated solid microneedles patch called Micro-Patch. It enables self-administration via three simple steps of press on, peel off and dispose. Their novelty lies in the development of aseptic process of formulating a solid dose of teriparatid and mounting it in a solid array.

Latch Medical, a company from Ireland has introduced solid coated microneedles, unique in their own way. Their uniqueness lies in the arrays of angled and interdigitated stainless steel microneedles which provide good anchoring on skin. These angled microneedles not only prevent skin deformation during insertion but also make the patch adhesive free.

Kindeva Drug Delivery, US, has commercialized both solid and hollow microneedle patches named solid microstructured transdermal systems (solid MTS) and hollow microstructured transdermal systems(h-MTS). The focus of these devices is predominantly towards cancer vaccines. One variant of the hollow microneedles delivers drugs to immune cell rich dermis. Instead of a patch, the device is in the form of a handheld applicator. The device structure can be said to be made up of three parts. The first part is for holding the device, the second part is where the drug reservoir is housed and the third is the hollow microneedle array part. The reservoir is refillable and the hollow microneedle array patch could be replaced as required.

The debioject microneedles developed at Debiotech, Switzerland [71] are examples of fine craftsmanship at microscale and show the advancement of expertise in hollow microneedle fabrication. They may be used individually or in an array with a length ranging from 350–900 µm. These needles can inject up to 500 µL in 5 minutes and work with viscous formulations as well. The ultra sharp geometry of the microneedles is obtained by varying the process parameters of DRIE.

3.9   CURRENT STATUS

The increasing importance of microneedle-based research could be understood from the fact that in 2020 microneedles were recognized as the top technology research area in World Economic Forum reports. An interesting and promising application of microneedles has been the use of the microneedle patch as a personal healthcare monitoring device which measures cholesterol and blood sugar through interstitial fluids [39]. Microneedle-based devices evolved to have components like heaters and sensor monitoring temperatures, humidity, glucose and pH [74].

The microneedle technologies which are in advanced clinical trial stages are insulin delivery, palmar hyperhidrosis, uveitis treatment, delivery of topical Lidocaine in oral cavity and migraine treatment [75].

3.10   CONCLUSION

This chapter discussed various aspects of microneedles. Their applications, types, structural materials etc. It gives the reader an overview of the world of microneedles. After having gained this insight, the user has to decide on which kind of microneedles will suit their purpose before they proceed with the design phase. The design of the microneedle system depends essentially on application and the kind of resources available among many factors. Table 3.2 lists the essential requirements of a microneedle-based product which helps in our choices while designing a drug delivery system based on microneedles.

TABLE 3.2
Microneedle Requirements

Parameter	Requirement	Choices available
Device Use	One time / prolonged time / use and throw / refillable / others	Patch / single device / multicomponent integrated system
Drug Delivery	Limited quantity of drug / sustained release / controlled flowrate / large quantity of drug / drug molecule size	Poke and patch / coated / dissolving / hydrogel based / hollow microneedles / others
Structural Material	Biocompatible, insertion into skin without breaking (high yield strength), easy regulatory compliances	Metals and alloys / polymers / others
Process	Batch manufacturable, high yield, scalable, repeatable, cost effective	Photolithography / droplet air-based blowing / deep reactive Ion etching (DRIE) / soft moulding / injection moulding / laser cutting / 3D printing / others
Plane	Ease of fabrication, design and application compliant	In plane / out of plane
Length	Should target dendritic cell rich dermis and well above nerve cells	350–900 μm
Single / Array	Larger fluid uptake area in skin, high delivery rate	Array
Structure	Insertion force should be less, pain perceived by patients should be less	Sharp bevelled tip, large side terminated lumen
Needle Spacing	Reducing bed of nails effect with skin deflection during microneedle insertion	250–500 μm
Width	High fracture force (large width), fewer tissues should be damaged (small width), pain perceived by patients should be less	160–800 μm
Lumen	Low fluid resistance of lumen (large lumen required) and clog free	>35 μm

After having made choices, one has to understand in detail the various fabrication processes for microneedles. The choice of the fabrication system is the make and break point for the success of the microneedle product. Many microneedles based products did not see the light of commercialization because their fabrication method was not scalable. The next chapter presents an overview of the different fabrication processes often used for microneedle fabrication.

REFERENCES

1. Ikeshoji, *Mosquito ecology – field sampling method*, Second Edition, Elsevier Applied Science, 1993.
2. M. D. R. Jones, S.J. Gubbins, Changes in the circadian flight activity of the mosquito *Anopheles gambiae* in relation to insemination, feeding and oviposition, *Physiol Entomol.* 3(3):213–220, 1978.
3. M. Macquity, *DK Inside guides – Amazing Bugs*, Dorling Kindersley Publishing, 15, 1996.
4. K. Tsuchiya, N. Nakanishi, Y. Uetsuji, E. Nakamachi. Development of blood extraction system for health monitoring system. *Biomed. Microdevices.* 7(4):347–353, Dec; 2005. DOI: 10.1007/s10544-005-6077-8
5. A. Kaushik, A.H. Hord, D.D. Denson, D.V. McAllister, S. Smitra, M.G. Allen, M.R. Prausnitz, Lack of pain associated with microfabricated microneedles, *Anesth. Analg.* 92:502–504, 2001.
6. R.K. Sivamani, B. Stoeber, G.C. Wu, H. Zhai, D. Liepmann, H. Maibach, Clinical microneedle injection of methyl nicotinate: stratum corneum penetration, *Skin Res. Tech.* 11(11):152–156, 2005.
7. K. van der Maaden, W. Jiskoot, J. Bouwstra, Microneedle technologies for (trans) dermal drug and vaccine delivery, *J. Control. Release.* 161(2), 2012.
8. B. Chen, J. Wei, F. Tay, Y.T. Wong, C. Iliescu, *Silicon Microneedles Array with Biodegradable Tips for Transdermal Drug Delivery*, EDA Publishing/DTIP, 2007. ISBN: 978-2-35500-000-3,.
9. M.S. Lhernould, Optimizing hollow microneedle arrays aimed at transdermal drug delivery, *Microsyst. Technol.* 19:1–8, 2013.
10. M. Belting, S. Sandgren, A. Wittrup, Nuclear delivery of macromolecules: Barriers and carriers, *Adv. Drug. Deliv. Rev.* 57:505–527, 2005.
11. D.B. das Olatunji, B. Al-Qallaf, Simulation based optimization of microneedle geometry to improve drug permeability in skin, *Proceeding of 7th Industrial Simulation Conference*, Loughborough, UK, 293–300, 2009.
12. R.F. Donnelly, T.R.R. Singh, D.I.J. Morrow, A.D. Woolfson, *Microneedle Mediated Transdermal and Intradermal Drug Delivery*, Wiley, 2012.
13. R F Donnelly, M T C McCrudden, A Z Alkilari, E Larranêta, E McAllister, A J Courtenay, et al., Hydrogel forming microneedles prepared from – super swelling polymers combined with hypophilised wafers for transdermal drug delivery, *PLOS ONE.*, vol. 9, e111547, 2014.
14. R F Donnelly, D I Morrow, M T C McCrudden, A Zaid Alkilari, E M Vincente–Pêrez, C O Mohony et al., Hydrogel forming and dissolving microneedles for enhanced delivery of photosensitizer and precursors, *Photochem. Photobiol.*, vol. 90, pp. 641–647, 2014.
15. Y C Kim, J H Park, and M R Prausnitz, Microneedles for drug and vaccine delivery, *Adv. Drug Deliv. Rev.*, vol. 64, pp. 1547–1568, 2012.
16. N Roxhed, P Griss, and G Stemme, Membrane-sealed hollow microneedles and related administration schemes for transdermal drug delivery, *Biomed. Microdevices.*, vol. 10, pp. 271–279, 2008.
17. K Ita, Transdermal delivery of drugs with microneedles – Potential and challenges, *Pharmaceutics*, vol.7, pp. 90–105, 2015.
18. M Ochoa, C Mousoulis, and B Ziaie, Polymeric microdevices for transdermal and subcutaneous drug delivery, *Adv. Drug Deliv. Rev.*, vol. 64, pp. 1603–1616, 2012.

19. E Larreneta, R E M Lutton, A D Woolfson, and R F Donelly, Miconeedle arrays as transdermal and intradermal drug delivery systems: Material Science, manufacturing and commercial development, *Mater. Sci. Eng.*, R104, 1–32, 2016.

20. R I Mahato and A Sarang, *Pharmaceutical dosage forms and drug delivery*, 2nd ed. CRC Press, pp. 423–425, 2011.

21. K E Peterson, Silicon as a mechanical material, *Proc. IEEE*, vol. 70, no. 5, pp. 420–457, 1982.

22. Á Cárcamo-Martínez, B Mallon, J Domínguez-Robles, L K Vora, Q K Anjani, R F Donnelly, Hollow microneedles: A perspective in biomedical applications, *Int. J. Pharm.*, vol. 599, 120455, ISSN 0378-5173, https://doi.org/10.1016/j.ijph arm.2021.120455, 2021.

23. J Yang, X Liu, Y Fu, Y Song. Recent advances of microneedles for biomedical applications: drug delivery and beyond. *Acta Pharm. Sin. B*. May; vol. 9, no. 3, pp. 469–483. doi: 10.1016/j.apsb.2019.03.007, 2019.

24. A F Moreira, C F Rodrigues, T A Jacinto, S P Miguel, E C Costa, I J Correia, Microneedle-based delivery devices for cancer therapy: A review, *Pharmacol. Res.*, vol. 148, p 104438, ISSN 1043-6618, https://doi.org/10.1016/j.phrs.2019.104438, 2019.

25. M. Shikida, T. Hasada and K. Sato, Fabrication of densely arrayed microneedles with flow channels by mechanical dicing and anisotropic wet etching, *J. Micromech. Microeng.*, vol. 16, pp. 1740–1747, 2006.

26. B Chen, J Wei, F Tay, Y T Wong, and C Iliescu, *Silicon microneedles array with biodegradable tips for transdermal drug delivery*, EDA Publishing/DTIP, ISBN: 978-2-35500-000-3, 2007

27. S A Coulman, A Anstey, C Gateley, A Morrissey, P McLoughlin, C Allendera, and J C Birchall, Microneedle mediated delivery of nanoparticles into human skin, *Int. J. Pharm.*, vol. 366, pp. 190–200, 2009.

28. Y Liu, P F Eng, O J Guy, K Roberts, H Ashraf, and N Knight, Advanced deep reactive-ion etching technology for hollow microneedles for transdermal blood sampling and drug delivery, *IET Nanobiotechnol.*, vol. 7, no. 2, pp. 59–62, 2013.

29. K B Vinayakumar, P. G Kulkarni, M M Nayak, N S Dinesh, G. M Hegde, S G Ramachandra and K Rajanna, A hollow stainless steel microneedle array to deliver insulin to a diabetic rat, *J. Micromech. Microeng.*, vol. 26, 065013, (9pp), 2016.

30. S Henry, D V Mc Allister, M G Allen, and M R Prausnitz, Microfabricated microneedles: a novel approach to transdermal drug delivery, *J. Harm. Sci.*, vol. 87, pp. 922–925, 1998.

31. K Kubo, N Tsukasa, M Uehara, Y Izumi, M Ogino, M Kitano, and T Sueda, Calcium and silicon from bioactive glass concerned with formation of nodules in periodontal ligament fibroblasts in vitro, *J. Oral Rehab.*, vol. 24, pp. 70–75, 1997.

32. J Amaral, V Pinto, T Costa, J Gaspar, R Ferreira, E Paz, *et al.*, Integration of TMR sensors in silicon microneedles for magnetic measurements of neurons, *IEEE Trans. Magn.*, vol. 49, pp. 3512–3515, 2013.

33. R Bhandari, S Negi, F Solzbacher, A novel mask-less method of fabricating high aspect ratio microneedles for blood sampling, 58th Electronic Components and Technology Conference, pp. 1306–1309, 2008.

34. W Zhang, J Gao, Q Zhu, et al., Penetration and distribution of PLGA nanoparticles in the human skin treated with microneedles. *Int. J. Pharm.*, vol. 402, no. 1–2, pp. 205–12, 2010.

35. Z Ding, F J Verbaan, B M Bivas, L Bungener, A Huckriede, D J Van den Berg, G Kersten, J A Bouwstra, Microneedle arrays for the transcutaneous immunization of diphtheria and influenza in BALB/c mice. *J. Contr. Rel.*, vol. 136, pp. 71–8, 2009.

36. S Sharma, K Hatware, P Bhadane, S Sindhikar, D K Mishra, Recent advances in microneedle composites for biomedical applications: Advanced drug delivery technologies. *Mater. Sci. Eng. C Mater. Biol. Appl.*, vol. 103, p. 109717. doi: 10.1016/j.msec.2019.05.002, 2019.

37. Martanto W, Davis S P, Holiday N R, Wang J, Gill H S, Prausnitz M R, Transdermal delivery of insulin using microneedles in vivo. *Pharm. Res.*, vol. 21, no. 6, pp. 947–52, doi: 10.1023/b:pham.0000029282.44140, 2004.

38. Gill H S, Prausnitz M R, Coated microneedles for transdermal delivery. *J. Control. Release.*, Feb 12, vol. 117, no. 2, pp. 227–37, doi: 10.1016/j.jconrel.2006.10.017, 2007.

39. Jin C Y, Han M H, Lee S S, Choi Y H, Mass producible and biocompatible microneedle patch and functional verification of its usefulness for transdermal drug delivery. *Biomed. Microdevices.*, Dec; vol. 11, no. 6, pp. 1195–203, doi: 10.1007/s10544-009-9337-1, 2009.

40. I Mansoor, Y Liu, U O Häfeli, B Stoeber, Arrays of hollow out-of-plane microneedles made by metal electrodeposition onto solvent cast conductive polymer structures, *J. Micromech. Microeng.*, **23** 085011, 2013.

41. D V McAllister, P M Wang, S P Davis, J H Park, P J Canatella, M G Allen, M R Prausnitz, Microfabricated needles for transdermal delivery of macromolecules and nanoparticles: fabrication methods and transport studies. *Proc. Natl. Acad. Sci. USA.*, Nov 25; vol. 100, no. 24, pp. 13755–60, doi: 10.1073/pnas.2331316100, 2003.

42. J H Park, Y K Yoon, S O Choi, M R Prausnitz, M G Allen. Tapered conical polymer microneedles fabricated using an integrated lens technique for transdermal drug delivery. *IEEE Trans Biomed Eng.* vol. 54, pp. 903–913, 2007.

43. R Luttge, E J W Berenschot, M J de Boer, D M Altpeter, E X Vrouwe, A van den Berg, and M Elwenspoek, Integrated lithographic moulding for microneedle-based devices, *IEEE ASME J. Microelectromech. Syst.*, vol. 16, no. 4, pp. 872–84, 2007.

44. P C Wang, B A Wester, S Rajaraman, S J Paik, S H Kim, M G Allen, Hollow polymer microneedle array fabricated by photolithography process combined with micromoulding technique. *Annu. Int. Conf. IEEE Eng. Med. Biol. Soc.*, pp. 7026–9, doi: 10.1109/IEMBS.2009.5333317, 2009.

45. B P Chaudhri, F Ceyssens, P De Moor, C Van Hoof and R Puers, A high aspect ratio SU-8 fabrication technique for hollow microneedles for transdermal drug delivery and blood extraction, *J. Micromech. Microeng.* vol. **20, p.** 064006, 10.1088/0960-1317/20/6/064006, 2010.

46. Liu, Shu & Wu, Dan & Quan, Ying-Shu & Kamiyama, Fumio & Kusamori, Kosuke & Katsumi, Hidemasa & Sakane, Toshiyasu & Yamamoto, Akira, Improvement of transdermal delivery of exendin-4 using novel tip-loaded microneedle arrays fabricated from hyaluronic acid. *Mol. Pharmaceutics.* 13. 10.1021/acs.molpharmaceut.5b00765, 2015.

47. BP Chaudhuri, Hoof Van C, R Puers, A novel method for monolithic fabrication of polymer microneedles on a platform for transdermal drug delivery. *Annu. Int. Conf. IEEE Eng. Med. Biol. Soc.*, pp. 156–9, doi: 10.1109/EMBC.2013.6609461, 2013.

48. H Takahashi, Y, Heo Jung, N, Arakawa *et al.*, Scalable fabrication of microneedle arrays via spatially controlled UV exposure. *Microsyst. Nanoeng.*, vol. **2**, p. 16049, https://doi.org/10.1038/micronano, 2016.

49. Z, Faraji Rad, R, Nordon, C, Anthony *et al.*, High-fidelity replication of thermoplastic microneedles with open microfluidic channels. *Microsyst. Nanoeng.*, vol. **3**, p. 17034, https://doi.org/10.1038/micronano, 2017.

50. R Mishra, T Kumar Maiti, T K Bhattacharyya, Development of SU-8 hollow microneedles on a silicon substrate with microfluidic interconnects for transdermal drug delivery, *J. Micromech. Microeng.* vol. **28**, p. 105017, 2018.

51. H Chiang, M Yu, A Aksit, W Wang, S Stern-Shavit, J W Kysar, A K Lalwani, 3D-printed microneedles create precise perforations in human round window membrane in situ. *Otol. Neurotol.* Feb; vol. 41, no. 2, pp. 277–284, doi: 10.1097/MAO.0000000000002480, 2020.

52. D Han, R S Morde, S Mariani, AA La Mattina, E Vignali, C Yang, G Barillaro, H Lee, 4D Printing of a bioinspired microneedle array with backward-facing barbs for enhanced tissue adhesion. *Adv. Funct. Mater.*, vol. 30, p. 1909197, 2020.

53. B D Ratner *et al. Biomaterial science: an introduction to materials in medicine*, Academic, New York, 1996.

54. Y Zheng, T Wang, X Tu, Y Huang, H Zhang, D Tan, W Jiang, S Cai, P Zhao, R Song, *et al.*, Gut microbiome affects the response to anti-PD-1 immunotherapy in patients with hepatocellular carcinoma. *J. Immunother. Cancer.* vol. 7, p. 193, doi: 10.1186/s40425-019-0650-9, 2019.

55. J A Mikszta, J P Dekker 3rd, N G Harvey, C H Dean, J M Brittingham, J Huang, V J Sullivan *et al.*, Microneedle-based intradermal delivery of the anthrax recombinant protective antigen vaccine. *Infect. Immun.*, Dec; vol. 74, no. 12, pp. 6806–6810. doi: 10.1128/IAI.01210-06, 2006.

56. M H Ling, M C Chen, Dissolving polymer microneedle patches for rapid and efficient transdermal delivery of insulin to diabetic rats. *Acta Biomater.* Nov; vol. 9, no. 11, pp. 8952–61, doi: 10.1016/j.actbio.2013.06.029, 2013.

57. W Yu, G Jiang, D Liu, L Li, H Chen, Y Liu, Q Huang, Z Tong, J Yao, X Kong, Fabrication of biodegradable composite microneedles based on calcium sulfate and gelatin for transdermal delivery of insulin. *Mater. Sci. Eng. C Mater. Biol. Appl.*, Feb 1; vol. 71, pp. 725–734, doi: 10.1016/j.msec.2016.10.063, 2017.

58. M C Chen, M H Ling, S J Kusuma, Poly-γ-glutamic acid microneedles with a supporting structure design as a potential tool for transdermal delivery of insulin. *Acta Biomater.*, Sep; vol. 24, pp. 106–16, doi: 10.1016/j.actbio.2015.06.021, 2015.

59. A Del Campo and C Greiner, SU-8: A photoresist for high aspect ratio and 3D submicron lithography, *J. Micromech. Microeng.*, vol. 17, R81–R95, 2007.

60. A J Teo, A Mishra, I Park, Y-J Kim, W T Park, and Y J Toon, Polymeric Biomaterials for medical implants and devices, *ACS Biomater. Sci. Eng.*, vol. 2, pp. 454–472, 2016.

61. F Clouet and M K Shi, Plasma surface modification of polymers: Relevance to adhesion, *J. App. Pol. Sci.*, vol. 46, pp. 1955, 1992.

62. J H Park, M G Allen, and M R Prausnitz, Biodegradable polymer microneedles: Fabrication, mechanics and transdermal drug delivery, *J. Control. Release*, vol. 104, pp. 51–66, 2005.

63. E I Vargha-Butler, E Kiss, C N C Lam, Z Keresztes, E Kálmán, L Zhang, et al., Wettability of biodegradable surfaces, *Colloid Polym. Sci.*, vol. 219, pp. 1160–1168, 2001.

64. C Ravindra, M Saraswati, G Sukanya, et al. Tensile and Thermal properties of Poly(vinyl) Pyrrolidone / Vanillin incorporated Polyvinyl alcohol films, *Res. J. Physical Sci.*, vol. 3, no. 8, pp. 1–6, 2015.

65. www.accudynetest.com.

66. D Wright *et al.*, Reusable, reversibly sealable parylene membranes for cell and protein patterning, *J. Biomed. Mater. Res–Part A*, vol. 85, pp. 530–538, 2008.

67. K Ahmed Saeed Al-Japairai, S Mahmood, S Hamed Almurisi, J Reddy Venugopal, A Rebhi Hilles, M Azmana, S Raman, Current trends in polymer microneedle for transdermal drug delivery. *Int. J. Pharm.*, Sep 25; vol. 587, pp. 119673. doi: 10.1016/j.ijpharm.2020.119673, 2020.

68. D Gan, T Xu, W Xing, M Wang, J Fang, K Wang, X Ge, C W Chan, F Ren, H Tan, X L,. Mussel-inspired dopamine oligomer intercalated tough and resilient gelatin

methacryloyl (GelMA) hydrogels for cartilage regeneration. *J. Mater. Chem. B.*, Mar 14; vol. 7, no. 10, pp. 1716–1725. doi: 10.1039/c8tb01664j, 2019.

69. www.azom.com/properties.aspx?ArticleID=1882
70. https://omnexus.specialchem.com/selection-guide/cyclic-olefin-copolymer/cyclic-ole fin-copolymer-properties
71. www.debiotech.com/debioject/
72. K, Márquez-Graña, S Bryan, C Vucen O'Sullivan, Development of a novel single-use microneedle design platform for increased patient compliance, *Procedia Manuf.*, vol. 13, pp. 1352–1359, ISSN 2351-9789, https://doi.org/10.1016/j.pro mfg.2017.09.114, 2017.
73. J Yang, X Liu, Y Fu, Y Song, Recent advances of microneedles for biomedical applications: drug delivery and beyond. *Acta Pharm. Sin. B.*, May; vol. 9, no. 3, pp. 469–483, doi: 10.1016/j.apsb.2019.03.007, 2019.
74. H Lee *et al.* Wearable/disposable sweat-based glucose monitoring device with multistage transdermal drug delivery module. *Sci. Adv.*, vol. 3, e1601314(2017), doi:10.1126/sciadv.1601314, 2017.
75. V Alimardani, S S Abolmaali, G Yousefi, Z Rahiminezhad, M Abedi, A Tamaddon, S Ahadian, Microneedle arrays combined with nanomedicine approaches for transdermal delivery of therapeutics. *J. Clin. Med.*, Jan 6; vol. 10, no. 2, p. 181, doi: 10.3390/jcm10020181, 2021.

4 Miniaturization Using Microfabrication Techniques

"There is plenty of room at the bottom"

—Nobel Laureate Richard Feynman

4.1 INTRODUCTION

Just as the depth of the ocean presents immense opportunities for exploration, in the same way there are many possibilities that wait to be explored at the micro and nanoscale. To make miniaturized devices, we need tools and methods that will get us there [1]. The earliest microfabrication techniques were used for integrated circuit fabrication. Now their horizon spreads to microfluidic devices, solar panels, displays, sensors and more. Microfabrication depends upon a countless number of factors. In microfabrication, we often come across two kinds of approach:

(i) Top-down approach, and
(ii) Bottom-up approach

In the top-down approach, we start from larger dimensions and then use microfabrication tools to bring it to the desired shape and characteristics. Alternatively, in a bottom-up approach, we build up our desired structure using atomic or molecular structures. This often employs self-assembly techniques. These approaches have been discussed in detail in literature. In this chapter we will be looking at different top-down microfabrication techniques which have been developed in recent years that enable us to make microdevices in general and microneedles in particular.

4.1 PHOTOLITHOGRAPHY

Photolithography is one of the most common top-down approaches used for microfabrication and dates back to as long ago as the 1820s, when the first photographic process was invented. During later periods, photolithography was commonly used for printed circuit boards and microprocessor fabrication. Now the photolithography process is extensively used for integrated circuit (IC) fabrication and MEMS devices. In photolithography, a light sensitive polymer is coated over a substrate which is then selectively exposed to ultraviolet light, using the mask. Upon exposure, the photosensitive polymer, known as resist, undergoes structural modifications, which is then selectively etched away using suitable chemicals (developer) – resulting in

DOI: 10.1201/9781003202264-4

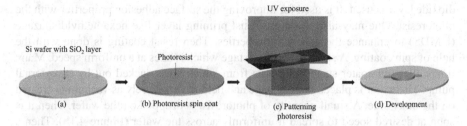

UV exposure

Si wafer with SiO₂ layer

Photoresist

(a) (b) Photoresist spin coat (c) Patterning photoresist (d) Development

FIGURE 4.1 Schematic of photolithography process.

pattern transfer. The etched resist regions are then filled with microneedle materials such as metals, ceramics and silicon. However, this process is overall costly and requires an advanced cleanroom facility with extended production time.

Essentially, the photolithography process consists of four basic process steps. These steps are shown in Figure 4.1.

(i) Clean
(ii) Coat
(iii) Expose
(iv) Develop

4.1.1 CLEAN

To start the photolithography process on an extremely clean wafer is a very important step as smoke, dust and bacteria destroy microdevices functionality. Most cleaning methods can be divided into two major groups: wet and dry methods. Dry cleaning processes use gas phase chemistry, and rely on chemical reactions required for wafer cleaning, as well as other techniques such as laser, aerosols and ozonated chemistries. Generally, dry cleaning technologies use less chemicals and thus are less hazardous for the environment but usually do not perform as well as wet methods, especially for particle removal. A commonly used cleaning technique involving silicon microfabrication is cleaning the silicon wafer using Piranha solution. A mixture of sulfuric acid (H_2SO_4) and hydrogen peroxide (H_2O_2) in the ratio 1:1 is used to clean the silicon wafer. This solution can be explosive. Hence care must be taken to mix the two ingredients. The wafer is lowered into the solution using a quartz stand and kept in the solution for about 30 minutes. Then the wafer is washed in deionized (DI) water and dried in the oven at around 150°C to drive away excess moisture. This is often called "dehydration bake". Another solution commonly used for cleaning wafers is acetone and isopropyl alcohol (IPA) solution (1:1) solution.

4.1.2 COAT

Photoresists are important components of the coat step. Photoresists are light sensitive materials which perform the two fold function of precise pattern formation and protecting the substrate. In the example shown in Figure 4.1, a silicon wafer with silicon dioxide layer is used. It is used for improving the surface adhesion properties with the photoresist. One may also use a pre-resist priming layer like hexamethyldisilazane (HMDS) to enhance the adhesion properties. Then resist coating is done with the help of spin coating. A spin coater has a stage which rotates at a uniform speed. Many times, the spin coater stage has a hole from which air is sucked out by an external pump. The wafer is placed on this stage and held there securely as vacuum is created on the other side. A small quantity of photoresist is dropped on the wafer. Then it is spun at desired speed to spread it uniformly across the wafer (Figure 4.1b). Then a soft bake step in an oven is followed to drive away excess solvent and improve the adhesion of photoresist on the wafer. Depending upon the change radiation causes to the photoresist structure, photoresists are classified as

(i) Positive photoresists, and
(ii) Negative photoresists

The irradiated regions in positive photoresists become more soluble in developer solutions as the ultraviolet light radiation breaks the existing bonds. On the contrary, in negative photoresists, the irradiated regions, new bonds are created and it is insoluble in the developer solution.

4.1.3 EXPOSE

The photoresist is then exposed to ultraviolet light using a mask (a glass having transparent and opaque area i.e. chrome layer corresponding to the desired pattern). The exposure time is extremely important as overexposure causes smaller resist patterns and underexposure leads to enlarged patterns. The typical emission spectrum of a mask aligner instrument with mercury (Hg) light source and without optical selective mirrors is g-line (436 nm) or i-line (365 nm). The optical absorption of most unexposed photoresists lies in the visible spectrum to near ultraviolet. Arriving at correct exposure time requires understanding the light source spectrum, measuring light intensity, and photoresist spectral range and sensitivity of the photoresist. In this step, the photoresist coated wafer is placed on the mask aligner stage and exposed to ultraviolet light as per calculated dose. In many cases, it is followed by examining the exposed area and post bake step and allowed to cool down.

4.1.4 DEVELOP

The purpose of development is to remove the exposed photoresist (in case of positive photoresist) and unexposed photoresist (in case of negative photoresist) and retain the final structure (Figure 4.1d). The substrate is dipped in the developer solution till the uncrosslinked photoresist gets detached from the substrate surface and dissolves in the developer solution. A careful visual examination and many times microscopic examination is required to carry on this step till the structures are fully developed and all the uncrosslinked photoresist is removed. The resist is then rinsed in appropriate solution like deionized water (DI) or iso-propyl alcohol (IPA).

4.1.5 ETCHING

Many times the photolithography step is accompanied by an etching step. Etching is selectively removing material from the bulk. There are two dominant etching techniques called wet etching and dry etching techniques. Wet etching of silicon is generally done by use of Potassium Hydroxide solution (KOH). This is anisotropic wet etching . Buffered Hydrofluoric acid may also be used. We shall study the wet and dry etching technique in more detail as another microfabrication technique in subsequent sections.

4.1.6 MICRONEEDLES BY UV PHOTOLITHOGRAPHY

Considering an example of fabrication of a microstructure like microneedles, we shall see the various photolithography steps involved. The proposed process flow for fabrication followed by wet etching technique is shown in Figure 4.2 [2]. These steps may vary from application to application but the major steps of photolithography remain same as those discussed above

4.1.6.1 Clean

To start with, a 2″ silicon wafer is used as a substrate and cleaned by the Piranha method (H_2SO_4 and H_2O_2 in the ratio 1:1) for 30 minutes and blown dry with nitrogen. Dehydration bake is carried out in the oven for 30 minutes to drive away the excess moisture from the wafer.

4.1.6.2 Oxidation

The oxidation of silicon is a diffusion process. Due to the relatively open structure of SiO_2, oxygen molecules or water molecules can diffuse through the growing SiO_2 layer and reach the silicon surface, where these molecules form SiO_2 layer as a mask layer for wet etching (Figure 4.2a). Thermal oxidation of silicon is achieved by heating the substrate to temperatures typically in the range of 900–1200 degrees C. The atmosphere in the furnace where oxidation takes place can either contain pure oxygen or water vapor. Both of these molecules diffuse easily through the growing SiO_2 layer at these high temperatures [3].

4.1.6.3 Photolithography and Wet Etch

The silicon wafer is patterned on both sides using UV photolithography. For this, the wafer is spincoated with negative photoresist (Figure 4.2b). After pre-baking it for 30 minutes in oven, it is exposed in a mask aligner under ultraviolet light for pattern containing square pattern for Potassium Hydroxide (KOH) etching of silicon. After the exposure, it is developed in developer and rinsed. It is postbaked giving resist pattern on one side of the wafer (Figure 4.2c). A similar procedure is carried out to pattern the second side of the wafer while aligning it to pattern on the other side of the wafer (Figure 4.2d–e). Once both sides of the wafer are patterned, then the silicon dioxide layer is removed from the unprotected region by etching with buffered Hydrofluoric acid (BHF) for 4–5 minutes. Piranha solution (H_2SO_4 and H_2O_2

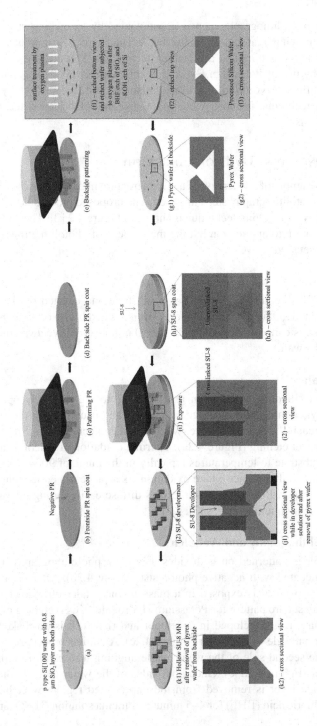

FIGURE 4.2 Suggested process steps for hollow microneedle fabrication by UV photolithography.

in the ratio 1:1) is used to remove the cross linked photoresist layer in 30 minutes. Then KOH (40% by weight) solution is prepared in a beaker and temperature was maintained at 70⁰C. Through holes are etched in silicon. The sample is then rinsed and dried. Figure 4.2 f1 and f2 show both the sides of the etched silicon wafer while Figure 4.2g shows the magnified schematic of the throughhole in wafer.

The wafer is then treated with oxygen plasma. Thick photoresist is spun on the substrate. The exposure step opens up the polymer photoresist ring which enhances the crosslinking efficiency. Post exposure bake is carried out (Figure 4.2d). This post exposure baking allows the photogenerated acid to diffuse in the photoresist and catalyze the reaction. Then the sample is cooled down gradually. It was developed in the developer solution (Figure 4.2e). After development, the sample is rinsed then slowly blown dry with nitrogen (Figure 4.2f). The resulting microneedle structures are shown in Figure 4.2f.

4.2 TWO PHOTON POLYMERIZATION-BASED MASKLESS LITHOGRAPHY

As indicated by the name, the resins are polymerized by absorbing two photons at a longer wavelength, usually in the near-infrared (NIR) spectral region (Figure 4.3). Three-dimensional laser lithography based on 2 PP is used for a variety of applications including micro-optics, photonics, and microfluidics [4]. It has several advantages over other conventional microfabrication techniques such as deep reactive-ion etching (DRIE), laser ablation, microstereolithography (μSL), drawing lithography, droplet born air blowing, and chemical isotropic etching, including the following:

(i)   The nonlinear response of the photoresists produces superior resolution
(ii) The method creates complex 3D structures directly from a CAD drawing, and
(iii) It allows fabrication of tall microstructures

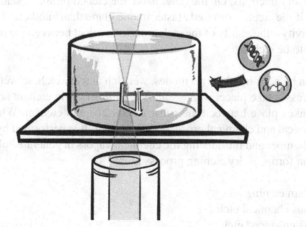

FIGURE 4.3   Principle of two photon polymerization done by a laser voxel resulting in 3D structure (courtesy www.microlight.fr/TPP.html accessed on 26.10.2021).

This is a novel technology that facilitates the fabrication of complex miniaturized structures that cannot be fabricated with established multistep manufacturing methods such as injection moulding, photolithography, and etching. Table 4.1 presents the review of the hollow Microneedles structures fabricated by two photon lithography.

4.3   SOFT MICROMOULDING

The soft moulding technique has been developed as an alternative to the photo-lithography techniques. It does not require an expensive cleanroom setup for microfabrication. This process generally requires the use of a mould as shown in Figure 4.4. The micro-moulding process consists of making replicates of the master mould. The mould is casted with a solution containing a polymer and active pharma-ceutical substances. Micro-moulding is considered a cost-effective method and is used for mass production. Micro-moulding is commonly used with polymer material for MN fabrication. The PDMS has several advantages in micro-moulding techniques such as low cost, ease of use, low surface energy, and thermal stability. The limitations associated with this technique are difficulties associated with controlling the depth of penetration, drug load capacity, and mechanical behaviour of the polymer.

4.4   ETCHING

Etching is required to selectively remove the areas defined by lithography or other microfabrication process or create structures for functional use. Generally, any etch process is characterized by its

(a) Etch rate – amount of material removed over a period of time.
(b) Uniformity – evenness of material removed from surface of wafer.
(c) Profile – the etched profile could be isotropic or anisotropic (Figure 4.5). An etching profile is said to be isotropic when the etching proceeds at an equal rate in all directions. On the other hand, the etched profile is said to be aniso-tropic if the etching proceeds faster in one plane than another.
(d) Selectivity – the ability of the etch process to select between the material to be or not to be etched.

Etching can be performed in two modes, wet etch and dry etch. In wet etch, purely chemical processes take place, where wafers are immersed in etchant solution. Here the reaction takes place between the substrate surface and etchant. Whereas in dry etch, both physical and chemical processes are present. It is performed by placing the wafer in the chamber and introducing the chemical vapors in generated plasma. Some of the common forms of dry etching processes are

1. Ion beam etching
2. Gaseous chemical etch
3. Plasma enhanced etch
4. Reactive ion etch
5. Deep reactive ion etch

TABLE 4.1
Review of Work Done in Fabrication of Hollow Microneedles by Two Photon Lithography

MN structures	Fabrication	Bioapplication	Materials	Device/femtosecond laser pulse source	Reference
Variable sizes of solid and hollow MNs with heights ranging from 375 to 750 μm and base 125 to 250 μm	MN directly fabricated through 2PP	MN arrays used to inject quantum dots into porcine skin	E-Shell 300 (acrylate-based polymer)	Ti:Sapphire laser (60 fs, 320 mW, 780 nm)	[5]
Hollow MNs with a 1450 μm height, 440 μm width, and 165 μm triangular bore	MN directly fabricated through 2PP integrated with microfluidic chip	Point-of-care MN device for detecting potassium	E-Shell 300 polymer	A Ti:Sapphire laser was used for 2PP at 800 nm, 150 fs, and 76 MHz	[6]
Hollow-bore pyramidal MNs with 1 mm height and 500 μm width	Photolithography, etching, and laser cutting processes	A lab-on-chip device for detecting proteins	E-Shell 300 polymer	A Ti:Sapphire laser was used for 2PP at 800 nm, 150 fs, and 76 MHz	[7]
250 and 300 μm Cone-shaped and pyramid-shaped structures	MN directly fabricated through 2PP	Point-of-care diagnostics	OrmoComp® (microresist technology, Germany)	(Amplitude Systems, Mikan), pulse duration of 300 fs and a repetition rate of 55 MHz at 515 nm	[8]
Up to 1300 μm high	Master MN arrays fabricated by 2PP. Silicone MN array moulds	Evaluated for their insertion in skin models	IP-S resist	Nanoscribe GmbH (780 nm, a pulse width of 150 fs, frequency of 40–100 MHz	[9]

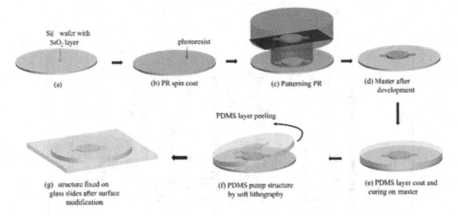

FIGURE 4.4 Process steps for soft micromoulding.

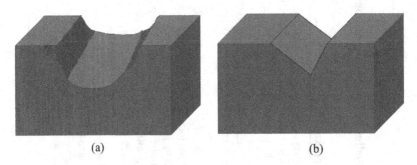

(a)   (b)

FIGURE 4.5 (a) Schematic of isotropic etch where a material gets removed uniformly from all directions and (b) shows anisotropic etch where the etching proceeds faster in one plane than another.

4.4.1 DEEP REACTIVE ION ETCHING PROCESS

The Deep Reactive Ion Etching (DRIE) is regarded as one of the powerful tools to fabricate silicon structures. It is best described as ion assisted etching. It is said to be a combination technology where in the equipment can control the process parameters like the density of reactive species near the substrate and control of ion current and energy. In DRIE, plasma is generated in low pressure chamber. Plasma is required in DRIE so that the electrons can dissociate input gas into atoms. The DRIE equipment consists of two sources connected to chamber and substrate as shown in Figure 4.6. The Inductively Coupled Plasma (ICP) source and the Radio Frequency (RF) source. The ICP source transfers the power coupled to the plasma with minimal voltage difference between the plasma and the wafer. The ICP source is responsible for controlling the ion flux reaching the substrate. The second power source is the RF power source which is capacitively coupled [10].

DRIE Chamber

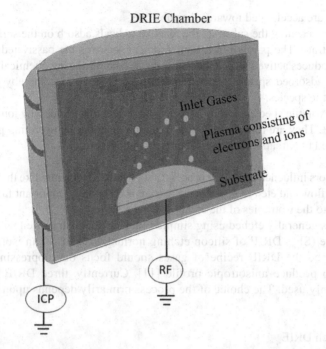

FIGURE 4.6 Schematic of a DRIE chamber which shows the connection for ICP and RF sources. The gases inlet, substrate and plasma consisting of both electrons and ions are shown in the chamber.

RF voltage is applied across metal plates to generate an oscillating electric field between them. The field accelerates electrons causing an ionization avalanche effect. To balance the plasma, a sheath layer providing an electric field perpendicular to the substrate is generated. This field is typically of the order of 13.56 MHz.

In the major bulk of the plasma, there are the same amount of positive and negative charges. Electrons, being lighter than plasma, can escape from plasma at a higher speed than ions. Once electrons are mostly depleted from boundaries i.e. the electrode region and the sample region, then this area will consist only of positive ions or neutral species. This is known as plasma sheath. It also creates positive plasma potential with regard to chamber walls. This sheet also accelerates ions and creates a sputtering effect on the sample.

Simply put, in the reactive etching chamber, following processes generally take place

(1) First the chamber is pumped down to low pressure conditions.
(2) As a next step, reactive gases are introduced in the chamber in a controlled way using mass flow controllers.
(3) The electrodes are powered to RF to drive the plasma discharge. This provides DC bias for electron acceleration which in turn allows the material placed on the electrode (substrate) to be exposed to ion bombardment.

(4) Ions are accelerated towards the substrate.
(5) Upon reaching the substrate the reactive radicals adsorb on the surface of the substrate. The parallel iron bombardment removes the passivated layer and reproduces active sites on the substrate surface for further chemical reactions.
(6) The adsorbed species and materials to be etched react, thereby producing volatile species.
(7) The volatile species desorb into gas phase, and this is aided by ion bombardment. They are removed from the reaction chamber by operating pumps and routed to a treatment unit before disposing of them.

Various factors influence the DRIE process and govern the etch rate like the ion power level, gases flow and etch temperature. Pressure is an equally important factor which contributes to the isotropies of the etch.

Silicon is generally etched using simple Fluorine chemistry often with Sulphur hexafluoride (SF_6). DRIE of silicon etching normally results in an isotropic etch profile. Hence the DRIE recipe of gases should focus on suppressing the lateral etch to produce anisotropic profile [11]. Currently, three DRIE processes are commonly used. The choice of the process primarily depends upon the user's application.

  i.   Bosch DRIE
  ii. Non-Bosch process
  iii. Cryogenic process

4.4.1.1 Bosch Process

The Bosch process was named after Robert Bosch GmbH, Germany, which patented the process. The process follows two alternate cycles

  (i) Etching with SF_6
  (ii) Depositing protective film on sidewall.

The etch profile obtained using the Bosch process has scallops due to the alternating cycle of etching and passivation, but this process is able to achieve a higher aspect ratio (Figure 4.7a). The etch rate is higher with higher selectivity as well.

4.4.1.2 Non-Bosch Process

The non-Bosch process is a similar process to the Bosch process where the bottom of the trench is etched with ions instead of SF_6.

4.4.1.3 Cryogenic Process

In the cryogenic etch process, lateral etch is suppressed by suppressing the chemical reaction between silicon and fluoride ions on the sidewalls by cooling the chamber to temperatures similar to liquid nitrogen typically -110°C. By using cryogenic etch one can obtain smooth sidewalls compared to Bosch process etch profile (Figure 4.7b).

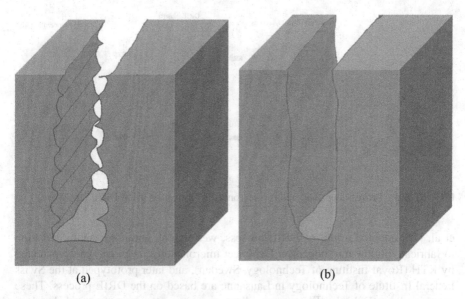

FIGURE 4.7   Etch profile obtained after Bosch process and cryogenic process-based etch.

Additionally, the process is economical to Bosch process as no alternate flow control cycles are required.

4.4.1.3   Microneedles Fabrication by DRIE Process

As an example of process flow, this section describes microneedle structure with the help of the DRIE process. The DRIE process has been used by researchers to fabricate mould as well as microneedle structures. Figure 4.8 shows the steps involved in hollow microneedle fabrication using DRIE.

Silicon wafer with silicon dioxide layer on both surfaces acts as substrate. This top layer is patterned using photolithography (Figure 4.8b) to define the mask protected sites. Majority of the wafer is exposed to etch. The substrate is kept in the DRIE chamber where it undergoes etching (Figure 4.8c), leading to solid microneedle shapes (Figure 4.8d). Then the substrate is reversed. Again the etching windows are defined by photolithography where the majority of the wafer is protected by SiO2 mask. The etching process defines the microneedle lumen this time (Figure 4.8e) yielding hollow microneedles as shown in magnified image (Figure 4.8f).

The major challenge in microneedle fabrication by DRIE is that the etching rate rapidly decreases if one is desirous of achieving high aspect ratio structures[12]. This happens as the ion flux is not able to reach the bottom of the structures and passivation layer removal does not take place. Many researchers adopted the pillar and drill method for silicon microneedle fabrication using DRIE. The gases commonly used for achieving anisotropic profile were C_4F_8, SF_6 and Argon. Doraiswamy et al. [13] used the Bosch process for creating high aspect ratio (30:1) microneedle structures with hollow lumen. To bring down the high operating cost of DRIE, many researchers also used a combination of wet etching and dry etching techniques [14,15]. Gardenier

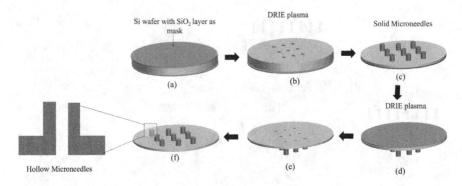

FIGURE 4.8   Process steps adopted for microneedle fabrication using DRIE technique.

et al. [16] adopted a mix of DRIE process, wet etching and conformal deposition to fabricate hollow microneedles. Debioject microneedles invented and co-patented by KTH (Royal Institute of Technology, Sweden), and later prototyped at the Swiss Federal Institute of Technology in Lausanne are based on the DRIE process. These microneedles (DebioJect™ microneedles) are currently manufactured by Leti (Grenoble, France) [17].

4.5   DRAWING LITHOGRAPHY

Drawing lithography is a technique that allows fabrication of 3-dimensional microneedles using 2-dimensional viscous liquid polymers. The entire field of drawing lithography is based upon the understanding of the glass transition of the polymer material for the microneedle. Glass transition region occurs between the solid and liquid states of a polymer [18]. It is the region for amorphous polymers (polymers not exhibiting any crystalline structure) at which increased molecular mobility due to increase in temperature results in dramatic changes in physical properties like hardness and elasticity. The glassy liquid on sub-strate offers increased processability and drawability. This technique also leads to reduced fabrication time. Taking the same example of microneedle fabrica-tion, it is also possible to load drugs in microneedles itself using this technique and there is no danger of chemical alteration or drug loss while loading. Various polymers being used in fabrication of microneedles using drawing lithography are carboxymethycellulose (CMC), Sodium Hyaluronate (HA), and polyvinyl-pyrrolidone (PVP).

This technique is used in ways in microneedle fabrication.

1.   Pillar-based drawing lithography
2.   Droplet air born (DAB) technique

In pillar-based drawing lithography, a planar substrate is coated with the polymer material and heated to its glass stage. Then a stage with pillars with diameter like the desired diameter of the microneedles is brought in contact with the viscous fluid and then lifted. The viscous fluid adhered to the pillars and then elongated when the upper stage moved upwards. At a point critical to elongation, the stage is held, and air blown to harden the polymer, thus resulting in solid microneedles.

The droplet airborne (DAB) technique is an advancement of pillar-based drawing lithography. In this technique the polymer droplet is shaped to a microneedle with the help of air blowing [19]. It offers advantages of easy control on drug volume to be loaded as it is relatively easier to control the drug droplet size by controlling volume and drug concentration.

The DAB technique, mostly used for solid microneedle fabrication, is described in detail below [19].

i.   **Droplet formation** – The dissolving polymer without the drug is dispensed on the substrate (Figure 4.9a). The precise control on droplet mass is achieved using an automated dispenser. Here the mass of the droplets on the patch is controlled by pressure and dispensing time. The next drug layer may be dispensed exactly on top of the previous droplets (Figure 4.9b).

ii. **Contact** – The upper stage is moved downwards till it is slightly in contact with the droplets at their positions and slightly deformed by it once the upper platform makes contact with them. The viscous polymer drops adhere to the upper plate (Figure 4.9c).

iii. **Microneedle length control** – After contacting the viscous polymer, the upper stage is moved upwards and held at a distance (Figure 4.9d). The upward movement elongates the viscous drop elongates. The optimum distance is calculated based on the required length of microneedles and is roughly twice the microneedle length.

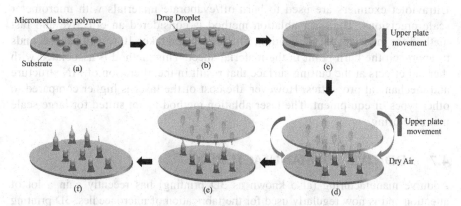

FIGURE 4.9 Process steps for droplet airborne technique (a) substrate loaded with microneedle base material droplet using dispenser, (b) drug droplets dispensed exactly aligned on the previous droplets, (c) upper platform moved downwards till the platform is in slight contact with the droplets, (d) upper plate is moved onwards while elongating the droplet till the point where the upper and lower drops are not detached and held there with dry air blown, (e) after hardening of structures, upper stage moved upwards again to break the structures at their neck, and (f) resulting microneedles.

iv. **Microneedle structure formation** – The stage is held at the optimum distance and then air is blown in between the plates. This step is performed to drive out excess solvents from polymers and help solidify the elongated droplets. Generally, an extended dumbbell kind of shape is formed between the plates (Figure 4.9d). The total length of the drawn structure depends on the drawing rate and time, hence offering control over microneedle length.

v. **Microneedle separation** – Once the polymer hardens, the upper plate is moved upwards quickly giving a slight pressure on the solidified structure. This jolt breaks the structures at their narrowest point (Figure 4.9e). Extra care should be taken during the design and optimization step that this narrowest part should be positioned exactly in the center of the solidified structure to have good yield of microneedles. After breaking the tapering and ultrasharp tip microneedles are formed on the substrate (Figure 4.9f).

Drawing Lithography has been mostly used to produce a variety of solid microneedles (Figure 4.9). An innovative way of hollow microneedle production using drawing lithography was shown by Hong et al. 2008 [20]. In this method, nickel was electroplated on the 3D solid structures of SU-8 photoresist fabricated by drawing lithography. After electroplating, the SU-8 mould structure was dissolved away in the SU-8 developer solution to yield high aspect ratio nickel hollow microneedles.

4.6 LASER ABLATION

Laser ablation focuses high energy, high intensity laser beams onto a substrate to create patterns/arrays of microneedles upon it. High energy lasers such as CO_2 Ultraviolet excimers are used to burn-off/evaporate materials with micrometer scale precisions. The laser ablation method is considered an effective and fast method for microneedle fabrication. The laser beam takes 10 to 100 nanoseconds to approach the burn point in the material sheet. This method is associated with thermal effects at the cutting surface that result in the alteration of MN structure and mechanical properties. However, the cost of the laser is higher compared to other types of equipment. The laser ablation method is not suited for large scale manufacturing.

4.7 ADDITIVE (3D) MANUFACTURING

Additive manufacturing (also known as 3D printing) has recently gain a lot of attention and is now regularly used for the fabrication of microneedles. 3D printing or additive manufacturing is the construction of a three-dimensional object from a CAD model or a digital 3D model. It can be done in a variety of processes in which material is deposited, joined or solidified under computer control, with material being added together (such as plastics, liquids or powder grains being fused), typically layer by layer. The biomedical device industry has seen a rapid rise of 3D printing technologies in tissue engineering implants in recent years. An advantage of using 3D printers to manufacture a microneedle array is the flexibility of design parameters and compressed lead times for processing.

CONCLUSION

In this chapter, we saw different approaches to microfabrication. It discussed essentially the top-down approaches which are commonly used for the fabrication of microneedles. Apart from traditional methods of photolithography, comparatively newer methods like dry etching, soft micromoulding and drawing lithography techniques were discussed.

REFERENCES

1. M.J. Madou, *Fundamentals of Microfabrication: The Science of Miniaturization*, Second Edition, CRC Press, 2017. https://doi.org/10.1201/9781482274004
2. R. Mishra, T. Maiti, T. Bhattacharyya, Development of SU-8 hollow microneedles on silicon substrate with microfluidic interconnects for transdermal drug delivery by IOP. *J.f Micromech. Microeng.* 28:105017, 2018.
3. B. Pramanick, MEMS based Silicon Microvalve and Micropump for Controlled Fluid Delivery in Micropropulsion Systems, PhD Thesis, Indian Institute of Technology Kharagpur, India, 2012.
4. W. Haske, V.W. Chen, J.M. Hales, W. Dong, S. Barlow, S.R. Marder, J.W. Perry, 65 nm feature sizes using visible wavelength 3-D multiphoton lithography. *Opt. Express.* 15:3426–3436, 2007.
5. S.D. Gittard, A. Ovsianikov, N.A. Monteiro-Riviere, J. Lusk, P. Morel, P. Minghetti, C. Lenardi, B.N. Chichkov, R.J. Narayan, Fabrication of polymer microneedles using a two-photon polymerization and micromolding process. *J. Diabetes Sci. Technol.* 3:304–311, 2009.
6. P.R. Miller, X. Xiao, I. Brener, D.B. Burckel, R. Narayan, R. Polsky. Microneedle-based transdermal sensor for on-chip potentiometric determination of K(+). *Adv. Healthc. Mater.* 3:876–881, 2014.
7. P. Miller, M. Moorman, R. Manginell, C. Ashlee, I. Brener, D. Wheeler, R. Narayan, R. Polsky. Towards an integrated microneedle total analysis chip for protein detection. *Electroanalysis.* 28:1305–1310, 2016.
8. A. Trautmann, G.L. Roth, B. Nujiqi, T. Walther, R. Hellmann, Towards a versatile point-of-care system combining femtosecond laser generated microfluidic channels and direct laser written microneedle arrays. *Microsyst. Nanoeng.* 5(6), 2019. DOI: 10.1038/s41378-019-0046-5.
9. A.S. Cordeiro, I.A. Tekko, M.H. Jomaa, L. Vora, E. McAlister, F. Volpe-Zanutto, M. Nethery, P.T. Baine, N. Mitchell, D.W. McNeill, R.F. Donnelly, Two-photon polymerisation 3D printing of microneedle array templates with versatile designs: application in the development of polymeric drug delivery systems. *Pharm. Res.* 37:174–174, 2020.
10. S. Tadigadapa, F. Lärmer, R. Ghodssi, P. Lin, (eds), Dry etching for micromachining applications. *MEMS Mater. Processes Handb.*, 403, MEMS Reference Shelf, Springer Science+Business Media, LLC (2011). DOI: 10.1007/978-0-387-47318-5_7, C _
11. www.samcointl.com/introduction-to-si-drie-silicon-deep-reactive-ion-etching/.
12. Y. Li, H. Zhang, R. Yang, Y. Laffitte, U. Schmill, W. Hu, M. Kaddoura, E.J.M. Blondeel, B. Cui. Fabrication of sharp silicon hollow microneedles by deep-reactive ion etching

towards minimally invasive diagnostics. *Microsyst Nanoeng*. 5:41, 2019. https://doi.org/10.1038/s41378-019-0077-y.

13. A. Doraiswamy, C. Jin, R.J. Narayan, P. Mageswaran, P. Mente, R. Modi, R. Auyeung, D.B. Chrisey, A. Ovsianikov, B. Chichkov, Two photon induced polymerization of organic-inorganic hybrid biomaterials for microstructured medical devices. *Acta Biomater*. 2:267–275, 2006.

14. N. Roxhed, T.C. Gasser, P. Griss, G. A. Holzapfel, G. Stemme, Penetration-enhanced ultrasharp microneedles and prediction on skin interaction for e_cient transdermal drug delivery. *J. Microelectromech. Syst*. 16:1429–1440, 2007.

15. Y.C. Kim, J.H. Park, M.R. Prausnitz, Microneedles for drug and vaccine delivery. *Adv. Drug Deliv. Rev*. 64:1547–1568, 2012.

16. J. Han, G.E. Gardeniers, R.L. Erwin, J.W. Berenschot, M.J. de Boer, S.Y. Yeshurun, M. Hefetz, R. van't Oever, A. van den Berg, Silicon micromachined hollow microneedles for transdermal liquid transport. *J. Microelectromech. Syst*. 12(6), 2003.

17. www.debiotech.com/debioject/, accessed on 01.12.2021.

18. K. Lee, H. Jung, Drawing lithography for microneedles: A review of fundamentals and biomedical applications. *Biomaterials*. 33(30):7309–7326, 2012. ISSN 0142-9612, https://doi.org/10.1016/j.biomaterials.2012.06.065

19. J.D. Kim, M. Kim, H. Yang, K. Lee, H. Jung, Droplet-born air blowing: Novel dissolving microneedle fabrication. *J. Control. Release*. 170(3):430–436, 2013. https://doi.org/10.1016/j.jconrel.2013.05.026

20. H.S. Hong, K.S. Han, K.J Byeon, H. Lee, K.W. Choi, Fabrication of sub-100 nm sized patterns on curved acryl substrate using a flexible stamp, *Jpn. J. Appl. Phys*. 47:3699, 2008. DOI: 10.1143/JJAP.47.3699

5 Design and Simulation of Microneedles

5.1 INTRODUCTION

Once an idea is born, the next steps are to design and validate it. The design is then analyzed from different aspects so that each parameter contributing in the design are optimized. It is useful to adopt a feasible design strategy in microelectromechanical systems (MEMS) by considering available fabrication techniques. Khalifa et al. (2010) [1] generalized that MEMS design starts with the following steps:

a) Defining overall system specification
b) Generation of comprehensive view of the system, and finally
c) Fabrication and characterization

In order to have a comprehensive view of the system as mentioned, we need to model the device. Modelling often refers to prediction of device behaviour based on the fundamental physical equations governing it. Modelling and simulation of MEMS-based devices may be further broken down into the following steps: [2]

a) Creation of device model
b) Device's physical behaviour-based model
c) Visualization and analysis of simulation-based results

Commercial mechanical engineering and finite element analysis software are useful for understanding and analyzing a lot of parameters like stress, displacement, diffusive transport and fluid velocity. Such a software is Comsol Multiphysics. This chapter first describes the mathematical analysis involved while considering a simple cylindrical shape of the microneedle. Based upon our understanding of the forces that act on a microneedle while it inserts into skin, the maximum bending and compressive forces acting on the microneedle are analyzed. Further, considering the same cylindrical structure, an attempt is made to understand the fluid flow through the microneedle with the help of the laminar flow physics governing it. Subsequently, simulation-based analysis is carried out to support the theoretical findings. Finite element-based simulation procedures for solving the structural mechanics and fluid flow problems for microneedles have been done particularly by taking an example of the hollow microneedle.

DOI: 10.1201/9781003202264-5

5.2 THEORETICAL ANALYSIS OF HOLLOW MICRONEEDLE

Microneedles have gained importance in the last two decades among various trans-dermal drug delivery techniques as they can deliver a variety of drugs through stratum corneum (SC), the outermost layer of skin, painlessly. They can be administered in a variety of ways which govern whether the needle structure shall be solid or hollow or the MN shall be made out of the drug itself. Hollow microneedles offer the advantages of offering a large volume or continuous flow of drugs with control overflow rate. In this work, SU-8 is selected as a potential material for microneedle fabrication based on high strength, biocompatibility, low cost, possibility of light induced poly-merization and compatibility with microelectronic industry processes. After material selection the design of the microneedle is crucial as it determines the fabrication steps. Analytical studies on microneedles have been addressed widely in literature and various factors of design like microneedle geometry, fluid mechanics inside the microneedle and fluid flow have been addressed [3,4,5]. In other works, analytical simulation is combined with software-based simulations to optimize the microneedle geometry and material strength [6,7,8].

5.2.1 STRUCTURAL ANALYSIS OF MICRONEEDLE

A good microneedle design requires a prior understanding of the microneedle mechanics when it inserts into skin. The following considerations are required while arriving at initial microneedle dimensions as have been discussed earlier too.

i.   The microneedle lumen should be small enough to damage fewer tissues but at the same time large enough (>35 μm) so that it is clog-free from body fluid materials. Hence the initial inner diameter of the microneedle is chosen as 40 μm.

ii. The outer radius of the needle should be such that it has high fracture force but at the same time damages fewer tissues. This parameter needs to be optimized but 100 μm is chosen as a starting value.

iii. The length of the microneedle should be such that it reaches dendritic cells rich dermis and at the same time does not prick the nerve cells i.e. ideally between 350–900 μm. The length requirement should also take into consid-eration the limitations imposed by the fabrication methods available. In the example that we have considered, photolithography and direct laser writing processes are the available options. We have already discussed that the pre-ferred structural materials for the structure given in the example is negative photoresist SU-8 (from IBM). In the fabrication process, in one of the steps, SU-8 has to be coated on the substrate using a spin coat process. The SU-8 photoresist comes in different viscosity versions as defined by its product number. We have considered the SU8-2150 which is a highly viscous version of SU-8 as compared to other commercial formulations. It yields a maximum thickness of 500–600 μm at 1000 rpm in a single spin coat [9]. Hence, we have chosen 500 μm as the microneedle length which could be obtained in a single coat. Another method that has been reported in literature is to convert

SU-8 to dry chips and then coat them on the substrate to achieve desired thickness [10].

iv. The spacing between microneedles in an array is an important factor which ensures effective microneedle penetration. If the needle-to-needle spacing is small, then the needles will not reach the targeted depth. Olatinji (2009) [7] suggested that for a microneedle array, the centre-to-centre spacing should be greater than the product of the base radius and aspect ratio and has a minimum value of 2. He also showed for microneedles (400 µm length and diameter 60 µm) that the stress at microneedle tip decreases substantially beyond 300 µm microneedle interspacing. Qallaf et al. (2008) [11] carried out simulation-based optimization of microneedle spacing and found that for both solid as well as hollow microneedles, the optimized aspect ratio (defined by the ratio of centre-to-centre spacing and microneedle radius) is 4. Hence for a hollow microneedle of external diameter 100 µm, the optimum microneedle spacing is 400–500 µm, out of which a microneedle spacing of 500 µm was chosen in the next example.

5.2.1 EXAMPLE 5.1 – CYLINDRICAL MICRONEEDLE

Table 5.1 shows the preliminary parameters considered for example 5.1 microneedle analysis.

A microneedle experiences various kinds of loads like axial and transversal load during its insertion and removal from the tissue. An axial force is a force that acts on the central axis of the structure under consideration. Stretching force and compressive forces are typical examples of axial forces. On the other hand, transversal force is a tangential force that acts on a structure in reaction to an angular acceleration. Transversal forces acting on microneedles may try to bend it. When a microneedle tries to puncture skin, resistive forces try to bend as well as compress it. The applied axial force should be greater than the skin resistance force. During insertion, possible failure of the microneedle may occur due to bending or buckling. When an applied load is greater than the yield strength of the material, failure occurs. Different kinds of forces acting on the chosen cylindrical microneedle are shown below.

5.2.1.1 Skin Resistive Force

In order to successfully penetrate human skin, the applied force should be greater than the resistive force. The force required for puncturing skin is given by

$$F_{Skin} = P_{Puncturing} A \qquad (5.1)$$

Where $P_{Puncturing}$ is the pressure exerted by the dermis while puncturing and A is the cross-sectional area of the microneedle tip. Table 5.2 lists the typical values and different constants considered in this section of analysis.

Typically $P_{Puncturing}$ is 3.18 MPa but after skin is punctured, resistive forces decline drastically and it drops to 1.6 MPa because of the underlying softer layer [12].

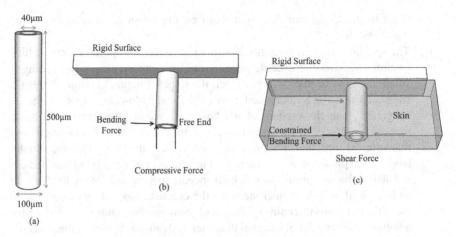

FIGURE 5.1   (a) Microneedle Schematics for example 5.1. Forces acting on microneedles (b) just as it enters the skin and (c) as it punctures and penetrates the skin.

TABLE 5.1
Initial Parameter Values Considered for Theoretical Analysis of Microneedles for Example 5.1

S/No.	Parameter	Symbol	Value
1	Structural material		SU-8
2	SU-8 Yield strength	σ_y	34 MPa
3	SU-8 Young's Modulus	E	4.4 GPa
4	Outer diameter	D_0	100 µm
5	Inner diameter	D_i	40 µm
6	Outer radius	R_o	50 µm
7	Inner radius	R_i	20 µm
8	Length	L	500 µm
9	Center to center spacing between microneedles		500 µm
10	Array		10X10

If we consider a simple hollow cylindrical geometry of microneedle as shown in Figure 5.1(a) then the cross-sectional area is given by

$$A = \pi \left(R_o^2 - R_i^2 \right) \tag{5.2}$$

Where R_0 and R_i are the outer and inner radius of a microneedle. The microneedle cross-sectional area is calculated to be 6597.34 µm². Substituting this value in equation 5.1, the skin piercing force for microneedles of given geometry is found to be 0.02 N for epidermis and 0.01 N for dermis. This is essentially the axial compressive force acting on the microneedle. For transverse forces, it is assumed here that it

TABLE 5.2
Parameter Values Considered Microneedles Analysis for Example 5.1

S/No.	Parameter	Symbol	Value
1	Pressure exerted by epidermis during puncturing	$P_{Puncturing}$	3.18 MPa
2	Pressure exerted by dermis during puncturing		1.6 MPa
3	Distance of vertical axis to outer edge	c	$D_0/2$
4	End condition constant	C'	0.25

is less than the axial forces, taken at $1/100^{th}$ of the axial force value (i.e. 0.0002 N for epidermis and 0.0001 N for dermis.

5.2.1.2 Maximum Compressive Force

The maximum compressive force that the microneedle can withstand without breaking is given by

$$F_{Compressive} = \sigma_y A \tag{5.3}$$

Where σ_y is the yield strength of the material. With yield strength of SU-8 as 34 MPa, the maximum compressive force for considered SU-8 microneedle geometry comes to be 0.2 N.

5.2.1.3 Maximum Bending Force

Initially, when the microneedle is inserted into the skin, the skin tissue may try to bend the microneedle. The maximum free bending force that a needle can withstand is given by maximum free bending force.

$$F_{Bending} = \frac{\sigma_y I}{cL} \tag{5.4}$$

Where I is the moment of inertia, and c is the distance of the vertical axis to the outer edge of section. In case of hollow cylindrical section, $c = D_0/2$ and I is given by

$$I = \frac{\pi \left(D_o^4 - D_i^4 \right)}{64} \tag{5.5}$$

I is calculated as 4783074.81 μm^4. Substituting this value in 5, we get maximum bending force for SU-8 microneedle as 0.014 N.

5.2.1.4 Maximum Buckling Force

Another loading condition that needs to be examined is buckling. Buckling may occur while the microneedle is being inserted into the skin. In order to calculate the maximum buckling force, the base of the microneedle is modelled as a fixed joint while the tip is modelled as a pivoted cylinder which results in an end condition constant C' of 0.25. Maximum Euler's buckling force is given by

$$F_{Buckling} = \frac{C\pi^2 EI}{L^2}$$

(5.6)

Where E is the Young's modulus (4.4 GPa for SU-8). Hence, we can calculate the maximum buckling force for SU-8 microneedles to be 0.18 N.

5.2.1.5 Maximum Constrained Bending Force

The forces which we have discussed above apply while inserting the microneedles into skin. Once the microneedle is inserted into skin, it can no longer move freely in lateral directions. Now it experiences a constrained bending force. The constrained bending force is given by

$$F_{ConstrainedBending} = \frac{2\sigma_y I}{cL}$$

(5.7)

The value of maximum constraint bending force is calculated to be 0.028 N.

5.2.1.6 Maximum Shear Force

After being fully inserted into the skin, there can be shear force acting on the microneedle due to the perpendicular movement between the base of the microneedle and microneedle itself. Maximum shear force which the microneedle can withstand is given by

$$F_{Shear} = \frac{\sigma_y A}{2}$$

(5.8)

Maximum shear force comes out to be 0.1 N.

The results obtained from our analysis for example 5.1 are mentioned in Table 5.3.

5.2.2 FLOW ANALYSIS OF MICRONEEDLE

Measurement and prediction of microneedles fluid dynamics is needed to design needles that balance between geometries small enough to avoid pain, sharp enough to easily insert into skin, and large enough to permit useful flow rates at reasonable pressures. Due to the small lumen of microneedles, delivery of large volumes through them becomes difficult. These needles can exhibit significant resistance to flow. This is the reason where in many cases, an array of microneedles is used to achieve larger flow rate than that achievable by a single needle. Also, the pressure drop required for fluid flow through microneedles depends on needle geometry, fluid viscosity and density. Contemporary micro pumping technologies used for drug delivery

TABLE 5.3
Different Forces Used in Simulation, Exerted as Resistance to Microneedle Penetration

S/No.	Parameter	Symbol	Value	Remarks
1	Skin resistive force epidermis (axial)	F_{Skin}	0.02 N	
2	Skin resistive force epidermis (transversal)	F_E	0.0002 N	Assumption that transversal forces are 1/100th of axial forces
3	Skin resistive force dermis (axial)	Fd_{Skin}	0.01 N	
4	Skin resistive force dermis (transversal)	F_d	0.0001 N	Assumption that transversal forces are 1/100th of axial forces

TABLE 5.4
Parameters Considered for Flow Analysis for Example 5.1

S/No.	Parameter	Symbol	Value
1	Fluid considered		Water
2	Dynamic viscosity of water	H	0.001 Pa.s
3	Kinematic viscosity	N	$1.004 \times 10^{-6} m^2/s$
4.	Pressure drop	p	1 kPa

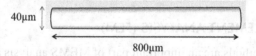

40μm

800μm

FIGURE 5.2 Dimensions considered for flow analysis of cylindrical microneedle for example 5.1.

applications typically operate in the range of 1–10 kPa [13]. Hence, a pressure drop of 1 kPa is considered initially for microneedles flow analysis.

5.2.2.1 Flowrate

Figure 5.2 shows the dimensions considered. Using Poiseuille's law of fluid flow for a cylinder, we can calculate the flowrate through MN.

$$Q = \frac{\pi D_i^4 p}{128 \eta L} \tag{5.9}$$

Where D_i is the inner diameter, p is the pressure difference between the outlet and inlet of microneedle, η is the dynamic viscosity and L is the length of the microneedle. Fluid considered is water.

Flowrate for a single microneedle is calculated to be 0.125 μL/s. When the microneedle is used as an array of 10X10 microneedle array, combined flowrate is 12.5 μL/s.

5.2.2.2 Reynold's Number

Another important parameter in flow analysis is the Reynold's Number. It is a dimensionless number which helps us to predict the flow pattern. The Reynolds Number is the ratio of inertial forces to viscous forces. High values of Reynold's Number indicate that viscous forces are smaller while low values of Reynold's Number indicates that the viscous forces dominate, and they should be taken into consideration. For a high Re > 4000, where inertial forces are dominant, the flow is known as turbulent flow. If the value of Re lies between 2100 and 4000, the flow is known as intermediate flow e.g., pipe flow, channel flow. When 1< Re < 2100 then the flow is said to be in laminar regime where viscous forces are dominant.

The Reynolds Number is given by:

$$\mathrm{Re} = \frac{QD_H}{v A}, \tag{5.10}$$

where D_H is the hydraulic diameter and A' is the cross-sectional area of lumen. For circular cross section $D_H = D_i$. For a microneedle of given geometry, the Reynold's Number is calculated as 3.984 which is very low. It indicates that flow through microneedle lumen shall be laminar.

The results from flow analysis in microneedles of chosen geometry is shown in table 5.5.

5.3   FINITE ELEMENT ANALYSIS (FEM)

Finite element methods are an important part of MEMS analysis and design. They allow us to predict how a design would react when subjected to the real world in terms of vibration, heat, fluid flow or other physical effects. It shows whether a product will work appropriately or will wear or break in actuality. Analytically, FEM is a computational technique to obtain approximations of the boundary value problems [14]. The boundary value problems are also called field problems. These field problems are representations of physical structures. Here, variables mean the field variables. They are governed by different differential equations. Displacement, temperature, velocity etc. may be considered to be variables depending upon the physical problem we are trying to analyze. To understand the contribution of FEM analysis, let us consider an example. Consider a two-dimensional system as shown below.

Consider the system shown in Figure 5.3 with a single field variable p(x,y). This variable is to be determined at point F(x,y). The important condition here is that the

TABLE 5.5
Results from Flow Analysis for Example 5.1

S/No.	Parameter	Value
1	Flowrate for single microneedle	0.125 μL/s
2	Flowrate for 10X10 microneedle	12.5 μL/s
3	Reynold's Number for microneedle lumen	3.984

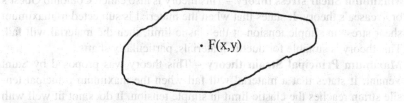

FIGURE 5.3 A general two dimensional domain of field variable p(x.y).

governing equation should be satisfied for the above shown system at each such point F(x,y). We can realize here that its solution will be a closed form of algebraic solution consisting of independent variables and it could be obtained with mathematical analysis. When we move to real-life designs, they are 3 dimensional and complex. This generates complex governing equations which have very less likelihood of converging solutions [15]. Hence, in FEM, a 3-dimensional object is broken down into a large number (thousands or even more) of finite elements, say, cubes. A computer program then calculates the behaviour of each cube depending upon the governing equations and boundary conditions. It then adds up the result of all these individual cubes and predicts behaviour of real-world problems. Some of the commonly used FEM software in the MEMS field are Ansys, Comsol Multiphysics, Autodesk simulation, MATLAB etc.

Considering the same example of a cylindrical hollow microneedle, we shall carry out FEM analysis for the same and investigate their degree of convergence with the theoretically calculated results. Comsol Multiphysics (version 5.2a) is used for this analysis.

5.3.1 STRUCTURAL ANALYSIS

The structural analysis is done to test the strength of a system. The structural analysis should act as a guide to our designs and instil confidence for their working. The stress analysis that we carry out depends on the stage of the product [16], namely:

i. The conceptual design of the product
ii. The detailed design
iii. Verification stage of the product

Stress analysis is required as most of the failures in engineering systems occur due to stress [17]. Various theories of failure that are used to analyze stress are:

i. **Maximum Principal stress theory** – It was proposed by Rankine and is valid for thin-walled tubes. It states that the maximum principal stress should not exceed the working stress of the material (the maximum allowable stress that a material or object will be subjected to while in work). It is generally applied to brittle materials under all loading conditions like biaxial or triaxial loading.

ii. **Maximum Shear stress theory** – This theory is also called Coulomb Guest's or Treasea's theory. It states that when the material is subjected to maximum shear stress in simple tension at the elastic limit, then the material will fail. This theory is suitable for ductile materials, particularly shafts.

iii. **Maximum Principal Strain theory** – This theory was proposed by Saint Venant. It states that a material will fail when the maximum principal tensile strain reaches the elastic limit in simple tension. It does not fit well with experimental results. It is appropriate for ductile and brittle materials under hydrostatic pressure.

iv. **Maximum strain energy theory** – It was proposed by Beltrani-Haigh. It states that the failure of material occurs when the total strain energy of the material reaches the total strain energy of the material at the elastic limit in simple tension. It is particularly applicable for pressure vessels.

v. **Maximum Distortion energy theory** – It was proposed by Von Mises-Henky. It states that when the shear strain energy per unit volume in the material under stress becomes equal to shear strain energy per unit volume at the elastic point in tension, then failure occurs. This is used in case of ductile materials which are subjected to pure shear.

The Von Mises stress is often used in determining whether a material will yield when subjected to a complex loading condition. This analysis is one of the closest matches to the experimental results. It is done by boiling the complex stress state down into a single scalar number that is compared to a metal's yield strength. In the case of microneedles and micropump, complex 3D loading is expected, hence we can use this criterion. In this approach, failure will occur if Von Mises stress induced in the material is greater than the yield strength of the material. In the next section we apply this model to microneedles of different geometries.

The examples considered here are presented with a view that the reader has a beginner level understanding of Comsol Multiphysics (for beginner's introduction of the software, one may refer to the Comsol Multiphysics manual). The default desktop of Comsol Multiphysics software is shown below [18].

The simulation is carried out generally as following steps:

i. Space – one starts building the model after choosing the space dimension
ii. Physics – adds physics interfaces which shall be acting on the model
iii. Study – adds studies to model
iv. Geometry (design) – using tools to create models

File menu   Customizable Quick   Ribbon tab and   Contextual tab   Ribbon group
             Access Toolbar   toolbar

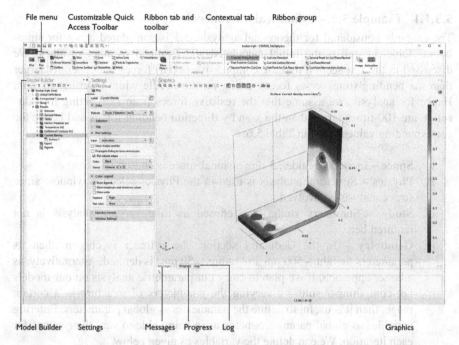

Model Builder   Settings   Messages   Progress Log   Graphics

FIGURE 5.4   The default COMSOL Desktop with its major windows in a widescreen layout.

Source: Courtesy Comsol Multiphysics manual website.

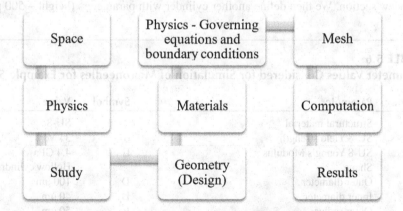

FIGURE 5.5   Sequence of modules general followed for Comsol Multiphysics.

v.   Materials – loads materials in model
vi.   Physics – governing equations and boundary conditions
vii. Mesh – discretized model into small units called mesh elements
viii. Computation – computes the results for model based on FEM analysis
ix.   Results – evaluates and visualizes results

5.3.1.1 Example 5.2 – Cylindrical Microneedle

The example considered for theoretical analysis, 5.1 is considered again for simulation. Table 5.6 relists the initial calculated and assumed parameter values. We discussed in the last section that the compressive (axial) forces are generally larger than the bending (transverse) forces acting on the needle while inserting in skin. Hence for analysis, we assume that the resistive forces from skin acting in z direction are 100 times larger than the x and y direction bending forces and hence the corresponding values taken in Table 5.6.

   i.   **Space** – In model builder, 3-dimensional space is chosen.
   ii. **Physics** – Structural analysis is used in the Physics selection window since stress analysis is involved.
   iii. **Study** – Stationary studies are chosen as time varying analysis is not required here.
   iv. **Geometry** – In the Geometry section, the cylinder is chosen, then its parameters (height – 500 µm and radius – 50 µm) is defined. Alternatively, as a better approach, if we plan to carry out parametric analysis on our models i.e. computing results by varying the parameters of structure in a certain range, then it is useful to define the parameters as global parameters. Entering variables as global parameters helps us in being able to vary their values for each iteration. We can define the variables as given below.

After entering these values in global parameters and recalling them as cylinder parameters in the geometry section, we build the cylinder. This creates a cylinder in the view section. We then define another cylinder with parameters (height – 500 µm

---

TABLE 5.6
Parameter Values Considered for Simulation of Microneedles for Example 5.2

S/No.	Parameter	Symbol	Value
1	Structural material		SU-8
2	SU-8 Yield strength	σ_y	34 MPa
3	SU-8 Young's Modulus	E	4.4 GPa
4	Shape		Hollow cylindrical
5	Outer diameter	D_0	100 µm
6	Inner diameter	D_i	40 µm
7	Outer radius	R_0	50 µm
8	Inner radius	R_i	20 µm
9	Length	L	500µm
10	Center to center spacing between microneedles		500 µm
11	Array		10X10
12	Skin resistance force in z direction	F_z	0.02 N
13	Skin resistance force in x direction	F_x	0.0002 N
14	Skin resistance force in y direction	F_y	0.0002 N

TABLE 5.7
Value of Variables Defined Under Global Parameters

S/No.	Variable Name	Value / expression	Description
1	Outer radius	50[um]	0.00005 m
2	Inner radius	20[um]	0.000002 m
3	Height	500[um]	0.0005 m

and radius – 20 μm). We can choose to perform Boolean operations on selected geometries. We perform the subtract operation by choosing outer cylinder as subtrahend and inner cylinder as minuend. On building this entity, we get the desired hollow cylindrical structure.

   v. **Materials** – We choose the SU-8 photoresist (solid out of plane) option from the material library and assign it to the cylindrical hollow structure. We use the linear elastic isotropic model for the microneedle. The material parameters that are required in this module are density, Young's modulus and Poisson's ratio.

  vi. **Physics** – In the structural mechanics module, we can analyze the mechanical behaviour of structures. It is an important module particularly for MEMS. On choosing this physics, the governing equations for this module are loaded onto the software.

The next step is to define the initial values and boundary conditions. The condition of fixed constraint is applied to the microneedle base attached to the substrate (Figure 5.6b). All other boundaries are free. Here we assume a changed scenario that the microneedle remains fixed while the skin moves towards the tip and not vice versa i.e. the resistive forces act on the microneedle tip while the microneedle base is fixed. It is important to understand here that first the tip of the needle comes in contact with skin. At this juncture, the skin resistive forces act on the needle tip. Once the needle penetrates skin, then the resistive forces are distributed throughout the needle body. This condition can be modelled using the features of point load, boundary load and body load in the structural mechanics module to aid in simulation. The changes which need to be applied while choosing a particular load is given below. Different results are obtained for different loading conditions where load is distributed as F_x, F_y and F_z as 0.0002 N, 0.0002 N and 0.02 N.

  (a) **Condition 1 (Point load)** – Skin resistive forces are applied to a point on the tip of the microneedle. This case is not relevant here since a simple cylindrical microneedle is considered, hence a better appropriation might be the boundary load which is discussed below.

  (b) **Condition 2 (Boundary load)** – Boundary load is applied to the tip of the microneedle. This is done by choosing the boundary load from the type of

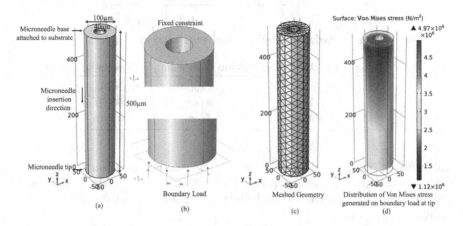

FIGURE 5.6 Different stages of cylindrical hollow microneedle simulation (a) structure geometry, (b) application of special conditions of fixed constraint on microneedle base and load (resistive force from skin while insertion) on microneedle tip, (c) meshed structure showing mesh elements and (d) 3-dimensional stress distribution of Von Mises stress throughout microneedle structure.

loads, selecting the tip of the needle where this force acts and finally, selecting total force from different kind of force options. The total force components are then defined as Fx = 0.01*Fz, Fy=0.01*Fz, and Fz=0.02 N.

(c) **Condition 3 (Body Load)** – Body load is chosen in the structural mechanics section and its values defined. It is chosen as an option from different kinds of loads in the similar fashion as the boundary load mentioned above.

vii. **Mesh** – The meshing sequence leads to the mesh. Before the "mesh build" command, we have to define the kind of mesh we want to build in our structure. One may choose from the different types of mesh elements and size or simply choose physics-controlled mesh and mesh size (rough, fine, finer) accordingly. The number of mesh elements decides the computation time. Choosing a higher number of mesh elements might result in very long calculation time. For this example, we choose physics-controlled mesh and select "build all". This discretizes the structure into 11821 tetrahedral elements, 2504 triangular elements and 344 edge elements for which results shall be calculated and used to predict failure (Figure 5.6c).

viii. **Computation** – One may proceed to compute the results by clicking on "compute". The software takes some time for computation and upon completion, the result is displayed in the result section.

ix.   **Result** – We obtain the 3-dimensional distribution of Von Mises stress throughout the needle structure. The colour chart in the figure in Comsol Multiphysics window defines the value of the Von Mises stress where colour blue (lighter shade in figure) indicates minimum value while red (darker shade in figure) indicates maximum value (Figure 5.6d). The maximum value of Von Mises stress is compared with the yield stress of structure material (SU-8) which is 34 MPa. Here we notice that the maximum Von Mises stress (both for boundary and body load is less than the yield strength of SU-8. This leads to the conclusion that the microneedle with given dimensions shall not break upon skin insertion and the geometry and material is good to be used for microneedle fabrication. Thus, we see that the simulation results are in agreement with the theoretical results for structural analysis as the same as the prediction from theoretical results. One may choose to represent the solution in a variety of ways visually or in table form from many options available.

x.   **Parametric Sweep**

There are a lot of variations one may choose to compute and analyze the results. One such step is the parametric sweep. In the study, one can include the parametric sweep and choose a parameter from defined global parameters and then define the range in which it is desired to be varied. The solution computed for this sweep may be visualized as a table or in the form of a graph. In the same example, we may keep the inner radius of the microneedle constant at 20 μm, and then vary the outer radius from 20 to 100 μm and compute the maximum Von Mises stress for the structure. The result is plotted as a line graph in Figure 5.7a. Alternatively, for the same structure, we may keep the outer radius of the needle constant at 50 μm and vary the inner radius to find the maximum Von Mises stress for the structure. It is plotted in the form of a point graph in Figure 5.7b. These results show low Von Mises stress (less than SU-8 yield strength) for microneedle dimension of our choice (i.e. length – 500 μm, outer radius – 50 μm and inner radius – 20 μm).

The results from structural analysis of cylindrical microneedles of given dimensions are given below.

Boundary load, maximum Von Mises stress – 4.97×10^6 N/m^2, region near base of microneedle

Body load, maximum Von Mises stress – 4.2×10^6 N/m^2, region near base of microneedle

5.3.1.2   Example 5.3 – Square Hollow Microneedles

One might also think of a square hollow design for a microneedle. Let us try to analyze one such structure. The parameters are indicated in Table 5.8. All other parameter data remain the same as in the last example.

i.   **Space** – In model builder, 3-dimensional space is chosen.
ii.   **Physics** – Structural analysis is used in the Physics selection window.
iii. **Study** – Stationary studies are chosen.
iv.   **Geometry** – First the outer cuboid (side – 100 μm) and then the inner cuboid (side – 40 μm) are constructed with height 500 μm. Then in a process similar to example 5.2, hollow square microneedle structure is created using Boolean subtract operation.

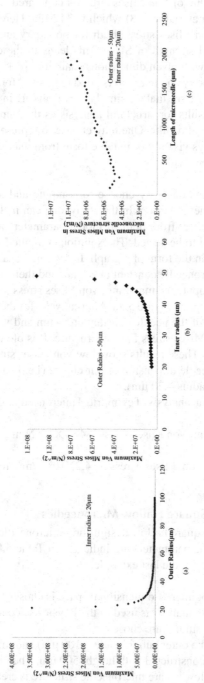

FIGURE 5.7 (a) Variation of Von Mises stress with outer radius (inner radius -20 μm), and (b) variation of Von Mises stress with inner radius (outer radius 50 μm) to show that choice of 50 μm outer radius and 20 μm inner radius generate low stress for the MN geometry.

TABLE 5.8
Parameter Values Considered for Simulation of Microneedles for Example 5.3

S/No	Parameter	Symbol	Value
1	Structural material		SU-8
2	Shape		Hollow square
3	Outer side	D_0	100 μm
4	Inner side	D_i	40 μm
5	Height	L	500 μm

v. **Materials** – We choose the SU-8 photoresist (solid out of plane) option from the material library and assign it to the hollow structure.

vi. **Physics** – On choosing this physics, the governing equations for this module are loaded onto the software.

vii. **Mesh** – For this example we choose physics-controlled mesh (fine) and select "build all". This discretizes the structure into 855 tetrahedral and 596 triangular, 136 edges and 16 vertex elements for which results shall be calculated and used to predict failure (Figure 5.8c).

viii. **Computation** – One may proceed to compute the results by clicking on "compute".

ix. **Result** – We obtain the 3-dimensional distribution of Von Mises stress throughout the needle structure. The maximum value of Von Mises stress is compared with the yield stress of structure material (SU-8) which is 34 MPa. Here we notice that the maximum Von Mises stress (both for boundary and body load is less than the yield strength of SU-8. This leads to the conclusion that the microneedle with given dimensions shall not break upon skin insertion and the geometry and material is good to be used for microneedle fabrication. Result can be summarized as

    a. Boundary load, maximum Von Mises stress – $3.92 \times 10^6 \, N/m^2$, region near base of MN

    b. Body load, maximum Von Mises stress – $3.26 \times 10^6 \, N/m^2$, region near base of microneedle

A good engineering practice calls for checking the feasibility of the chosen geometry by the available microfabrication techniques. One having preliminary knowledge of the photolithography process would understand instantaneously that such kind of sharp corners are not possible by photolithography or soft moulding techniques. The corners tend to round up at the sharp corners. There are some techniques of edge compensation available that may be considered if application calls for sharp edges.

5.3.1.3   Example 5.4 – Triangular Microneedles

Another possible design might be a triangular hollow design for microneedles. The process steps are given below. The parameters are indicated in Table 5.9. All other parameter data remain the same as in the last example.

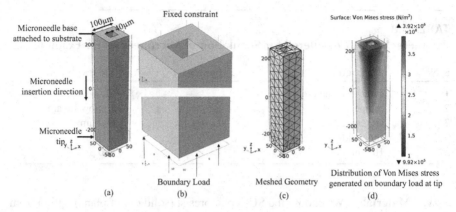

FIGURE 5.8 Different stages of square hollow microneedle simulation (a) structure geometry, (b) application of special conditions of fixed constraint on microneedle base and load (resistive force from skin while insertion) on microneedle tip, (c) meshed structure showing mesh elements and (d) 3-dimensional stress distribution of Von Mises stress throughout microneedle structure.

TABLE 5.9
Parameter Values Considered for Simulation of Microneedles for Example 5.4

S/No.	Parameter	Symbol	Value
1	Shape		Hollow triangular
2	Outer side	D_0	100 μm
3	Inner side	D_i	40 μm
4	Height	L	500 μm

i.   **Space** – In model builder, 3-dimensional space is chosen.
ii. **Physics** – Structural analysis is used in the Physics selection window.
iii. **Study** – Stationary studies are chosen.
iv. **Geometry** – For a triangular microneedle geometry, we shall construct a 2-dimensional base first and extrude it in the z plane to form a triangular microneedle. For this, the xy workplane is selected. In plane geometry, outer Bezier polygon (triangle) is made with coordinates (-50,0), (50,0) and (0, 86.6) and inner triangle is made with coordinates (-20,20), (20,20) and (0, 54.64). Then the inner triangle is subtracted from the outer triangle, yielding the desired triangular base. This structure is extruded 500 μm in z direction.
v.   **Materials** – We choose the SU-8 photoresist (solid out of plane) option from the material library and assign it to the hollow structure.
vi. **Physics** – On choosing this physics, the governing equations for this module are loaded onto the software.

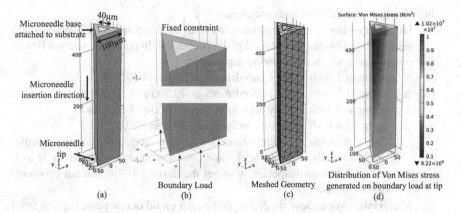

FIGURE 5.9 Different stages of triangular hollow microneedle simulation (a) structure geometry, (b) application of special conditions of fixed constraint on microneedle base and load (resistive force from skin while insertion) on microneedle tip, (c) meshed structure showing mesh elements and (d) 3-dimensional stress distribution of Von Mises stress throughout microneedle structure.

    vii.   **Mesh** – For this example we choose physics-controlled mesh (fine) and select "build all". This discretizes the structure into 968 tetrahedral, 668 triangular, 138 edge and 12 vertex elements for which results shall be calculated and used to predict failure (Figure 5.9c).

   viii.   **Computation** – One may proceed to compute the results by clicking on "compute".

    ix.   **Result** – We obtain the 3-dimensional distribution of Von Mises stress throughout the needle structure. The maximum value of Von Mises stress is compared with the yield stress of structure material (SU-8) which is 34 MPa. Here we notice that the maximum Von Mises stress (both for boundary and body load is less than the yield strength of SU-8. Maximum stress occurs at the tip of the microneedle. This leads to the conclusion that the microneedle with given dimensions shall not break upon skin insertion and the geometry and material is good to be used for microneedle fabrication. Result is summarized as follows:

        a.   Boundary load, maximum Von Mises stress – 1.04×10^7 N/m^2, region near base of microneedle

        b.   Body load, maximum Von Mises stress – 8.55×10^6 N/m^2, region near base of microneedle

5.3.1.4   Example 5.5 – Tapering Hollow Microneedle

The next design that we shall consider is tapering hollow microneedle. The process steps are given below.

    i.   **Space** – In model builder, 3-dimensional space is chosen.

    ii.   **Physics** – Structural analysis is used in the Physics selection window.

iii. **Study** – Stationary studies are chosen.
iv. **Geometry** – For a tapered geometry, we shall have to construct a 2-dimensional section first. For this, the yz workplane is selected. In plane geometry, a Bezier polygon is made with following segments:
   a. segment 1 with x and y coordinates as 20,0 and 40,500,
   b. segment 2 with x and y coordinates as 40,500 and 70,500
   c. segment 3 with x and y coordinates as 70,500 and 50,0 and
   d. segment 4 with x and y coordinates as 50,0 and 20,0
      This constructs a 2-dimensional lengthwise section of the microneedle. We can use the revolve workplane option with start and end angle as 0 and 360 degrees respectively to get the desired 3-dimensional geometry of the microneedle.
v. **Materials** – We choose the SU-8 photoresist (solid out of plane) option from the material library and assign it to the hollow structure.
vi. **Physics** – On choosing this physics, the governing equations for this module are loaded onto the software. The fixed constraint condition is applied to the base of the microneedle while the skin resistive forces act on the sharp tip of the microneedle during the course of insertion. This sharp tip also accounts for less stress and hence easy penetration of microneedles which shall be substantiated by our simulation results. In this example, too, we shall consider the cases where boundary and the body load act on the needle.
vii. **Mesh** – For this example we choose physics-controlled mesh (fine) and select "build all". This discretizes the structure into 8407 tetrahedral, 2618 triangular, 333 edge and 16 vertex elements for which results shall be calculated and used to predict failure (Figure 5.10c).
viii. **Computation** – One may proceed to compute the results by clicking on "compute".

We obtain the 3-dimensional distribution of Von Mises stress throughout the needle structure. The maximum value of Von Mises stress is compared with the yield stress of structure material (SU-8) which is 34 MPa. Here we notice that the maximum Von Mises stress (both for boundary and body load is less than the yield strength of

TABLE 5.10
Parameter Values Considered for Simulation of Microneedles for Example 5.5

S/No.	Parameter	Symbol	Value
1	Structural material		SU-8
2	Shape		Hollow tapering
3	Base outer diameter	D_{bo}	140 µm
4	Base inner diameter	D_{bi}	100 µm
5	Tip outer diameter	D_{to}	100 µm
6	Tip inner diameter	D_{ti}	40 µm
7	Height	L	500 µm

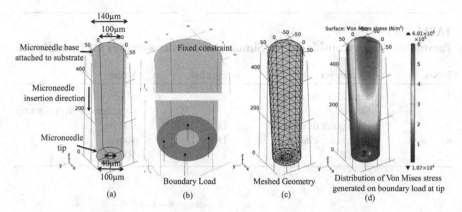

FIGURE 5.10 Different stages of tapering hollow microneedle simulation (a) structure geometry, (b) application of special conditions of fixed constraint on microneedle base and load (resistive force from skin while insertion) on microneedle tip, (c) meshed structure showing mesh elements and (d) 3-dimensional stress distribution of Von Mises stress throughout microneedle structure.

SU-8. This leads to the conclusion that the microneedle with given dimensions shall not break upon skin insertion. This kind of microneedle structure encounters very low insertion forces, and hence it is a very attractive option for microneedles. When the geometry is considered from the fabrication point of view, then photolithography processes do not allow a good control on tapering of the needle structure to yield the desired structure. Using 3D printing or two photon lithography processes leads to low throughput. Injection micromoulding is a good choice for tapering solid microneedles but the design of mould for hollow tapering structure becomes very intricate and may lead to damage of weak mould parts during micromoulding operation. Hence fabrication feasibility is an important aspect to be considered while picking up any microneedle geometry.

RESULT

Boundary load, maximum Von Mises stress – 6.01×10^5 N/m^2, region near base of microneedle

Body load, maximum Von Mises stress – 2.81×10^6 N/m^2, region near base of microneedle

5.3.1.4   Example 5.6 – Slant Tip Microneedle

One of the special designs that we shall consider next is the slant tip microneedle. This shape imitates the hypodermic syringe tips.

The process steps are given below.

i.   **Space** – In model builder, 3-dimensional space is chosen.
ii. **Physics** – Structural analysis is used in the Physics selection window.

TABLE 5.11

Parameter Values Considered for Simulation of Microneedles for Example 5.6

S/No.	Parameter	Symbol	Value
1	Structural material		SU-8
2	Shape		Cylinderical with slant tip
3	Base outer diameter	D_o	100 μm
4	Base inner diameter	D_i	40 μm
5	Height	L	500 μm

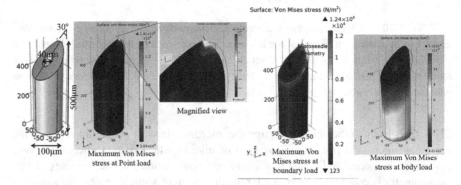

FIGURE 5.11 Microneedle structure geometry showing maximum Von Mises stress for slanted geometry in the first subsection where the microneedle has very high stress at the tip which might break the tip (as shown in inset in 2nd subsection). The load decreases as the microneedle penetrates as shown by the 3rd picture and is even lesser when the microneedle has completely penetrated the skin (4th picture). Cylindrical geometry is chosen for microneedles due to ease of fabrication and no issue of tip breakage.

   iii. **Study** – Stationary studies are chosen.

   iv. **Geometry** – First the outer triangle (side – 100 μm) and then the inner triangle (side – 40 μm) are constructed with height 500 μm. Then in a process similar to example 5.2, hollow triangular microneedle structure is created using Boolean subtract operation.

   v. **Materials** – We choose the SU-8 photoresist (solid out of plane) option from the material library and assign it to the hollow structure.

   vi. **Physics** – On choosing this physics, the governing equations for this module are loaded onto the software. The fixed constraint condition is applied to the base of the microneedle while the skin resistive forces act on the sharp tip of the microneedle during the course of insertion. This sharp tip also accounts for less stress and hence easy penetration of microneedles which shall be substantiated by our simulation results. In this example, too, we shall consider the cases where boundary and the body load act on the needle.

vii. **Mesh** – For this example we choose physics controlled mesh (fine) and select "build all". This discretizes the structure into free tetrahedral elements for which results shall be calculated and used to predict failure.

viii. **Computation** – One may proceed to compute the results by clicking on "compute".

ix. **Result** – We obtain the 3-dimensional distribution of Von Mises stress throughout the needle structure. The maximum value of Von Mises stress is compared with the yield stress of structure material (SU-8) which is 34 MPa. Here we notice that the maximum Von Mises stress (both for boundary and body load is less than the yield strength of SU-8. This leads to the conclusion that the microneedle with given dimensions shall not break upon skin insertion and the geometry and material is good to be used for microneedle fabrication.

This kind of microneedle structure encounters very low insertion forces, and hence it is a very attractive option for microneedles. When the geometry is considered from the fabrication point of view, then photolithography processes do not allow a good control on tapering of the needle structure to yield the desired structure. Using 3D printing or two photon lithography processes leads to low throughput. Injection micromoulding is a good choice for tapering solid microneedles but the design of mould for hollow tapering structure becomes very intricate and may lead to damage of weak mould parts during micromoulding operation. Hence fabrication feasibility is an important aspect to be considered while picking up any microneedle geometry.

5.3.2 Fluidic Analysis

For simulation based fluidic analysis of microneedle array(10X10), the 2-d view of the fluidic connectivity of the drug delivery system is shown in Figure 5.12a. The dimensions of the reservoir and the substrate are chosen keeping in view the dimensions achievable by the microfabrication processes. Using laminar flow model, velocity variations along a cut line (Figure 5.12a), extending from drug reservoir inlet to MN outlet, is calculated and results shown in Figure 3.6(b). Water velocity at the microneedle outlet is 0.3 m/s. Similarly, pressure drop is calculated along the same line through the length of the reservoir and microneedle. The 1 kPa pressure difference at the drug reservoir inlet continuously decreases at microneedle outlet (Figure 5.13a). The Reynold's Number for the fluid flow is found to be 0.56 indicating viscous laminar flow (Figure 5.13b).

5.3.3 Drug Diffusion-based Studies

For modelling the fluid flow through the blood veins, it is important to understand the pharmacokinetics of transdermal drug delivery. When a microneedle filled with drug is inserted into the skin, a concentration gradient develops, as a result of which drug starts to move down the skin layers. When it reaches the capillary vasculature, it is absorbed there and then is transported into the systemic circulation. Sometimes the microneedle may reach to the blood vein directly to deliver the drug. The veins are of order of 100 μm in size.

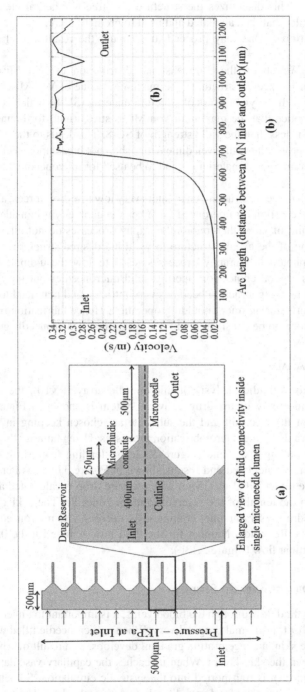

FIGURE 5.12 (a) pressure variation along the chamber and microneedle. Inlet pressure is 1 KPa, (b) variation of Reynold's Number along the cutline with maximum Reynold's Number as 0.56.

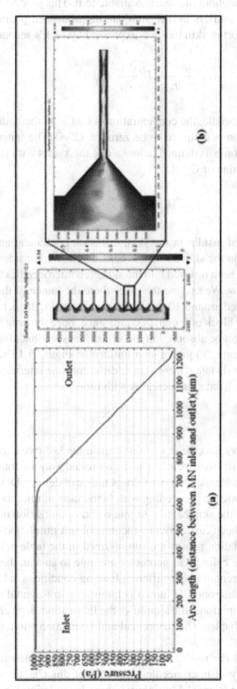

FIGURE 5.13 (a) 2-dimensional view of the fluidic path of the drug delivery system consisting of 10 microneedles connected to the reservoir chamber through microfluidic ports showing the dimensions considered for simulation studies. (b) Variation of velocity within the fluidic chamber and microneedle along the cutline.

Diffusion occurs by random kinetic movement of molecules and tends to spread any substance evenly throughout the space available to it. This process is not affected by drugs directly but the barriers to diffusion can be influenced pharmacologically. The transient drug transport in skin tissue is expressed by Fick's second law as

$$\frac{\partial C''}{\partial t} = D' \frac{\partial^2 C''}{\partial x^2} \tag{5.11}$$

At the tip of the microneedle, the concentration is $C=C_0$. At the bottom of the epidermis, drug concentration is assumed to be zero i.e. $C''=0$. The initial condition is given by a bell-shaped profile as defined below along the x-axis with its maximum at $x=0$ and corresponding value of $C''= C_0$

$$C''(t_0) = C_0 e^{-ax^2} \tag{5.12}$$

For modelling diffusion of insulin in the skin, diffusion coefficient and peak initial concentration of insulin on the skin surface is considered. We consider the tip of the microneedles which have been inserted into the skin. Ten microneedles are considered for the 2-dimensional view. We assume the case where the lumen of the microneedle is 40 μm and has penetrated around 10 μm in the skin layer (Figure 5.14). The skin is modelled as a rectangular block of length 500 μm, since, once the drug has penetrated in skin to this depth, it shall be absorbed in the veins. It is seen that after 600 seconds insulin has diffused through 250 μm of skin uniformly. Figure 5.15 shows the concentration profile with the distance from skin at different time intervals. The concentration at a given distance from skin increases with time.

5.4   CONCLUSION

Hollow microneedle arrays have been sustained since the last two decades and hold a lot of promise as an effective, painless and controlled way of transdermal drug delivery because of the immense variety of drugs they can deliver. Design and simulation are quintessential prior to proceeding with fabrication of microneedles in order to successfully overcome the skin barrier. In this work, optimization of the design of hollow microneedles has been carried out by theoretical and simulation-based studies. The theoretical and simulation results are summarized in the table below. The SU-8 microneedles with simple cylindrical geometry are able to sustain the skin resistive forces since their yield strength under different loading conditions is higher than the skin forces. The flow in microneedle lumen is laminar with Reynold's Number 3.9. The choice of cylindrical microneedle simplifies the fabrication steps greatly which is shown in the following chapter. The theoretical and simulation studies-based results are tabulated below.

This chapter equipped the readers with design, theoretical analysis and simulation of different geometry of microneedles. The aim of this chapter was to present an exhaustive exercise for Comsol Multiphysics based simulation for microneedles.

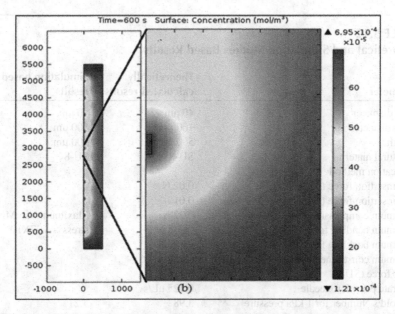

FIGURE 5.14 Drug diffusion profile for ten microneedles when inserted in skin and enlarged view for diffusion profile for a single microneedle in inset.

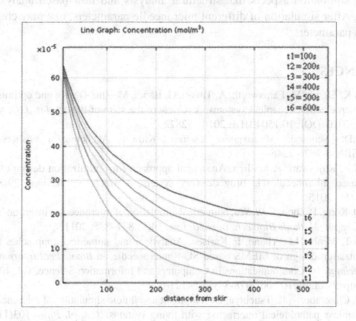

FIGURE 5.15 Variation of drug concentration profile with distance in skin at different time intervals.

TABLE 5.12
Theoretical and Simulation Studies Based Results

Parameter	Theoretically calculated results	Simulation based results
Inner diameter	40 μm	40 μm
Outer diameter	100 μm	100 μm
Length	500 μm	500 μm
Structural material	SU-8	SU-8
Fabrication method		
Skin insertion force (dermis)	0.02 N	
Skin insertion force (epidermis)	0.01	
Maximum compressive force (SU-8)	0.22 N	Maximum Von Mises
Maximum bending force (SU-8)	0.014 N	stress – 4.97X10^6
Maximum buckling force (SU-8)	0.207 N	
Maximum constrained bending force (SU-8)	0.028 N	
Shear force (SU-8)	0.1 N	
Flowrate per microneedle	0.125 μL/s	
Reynold's Number for 1 kPa pressure	3.98	3.21

Different simulation aspects like structural analysis and flow-based analysis were taken up. After simulation of different microneedle parameters, one may choose the optimum parameters.

REFERENCES

1. B K. Elhadj, M. Farnsworth, A. Tiwari, G. Bandi, M. Zhu, Design and optimisation of microelectromechanical systems: A review of the state-of-the-art. *Int. J. Design Eng.* 3, 2010. DOI: 10.1504/IJDE.2010.032822
2. S.D. Senturia, *Microsystem Design*. Kluwer Academic Publishers, 2001. ISBN-0-7923-7246-8.
3. N.R. Rajeswari, P. Malliga, Analytical approach for optimization design of MEMS based microneedles in drug delivery system. *J. Mech. Sc. Technol.* 29(8):3405–3415, 2015.
4. Q. Kong, P. Zhou, C.W. Wu, Numerical simulation of microneedles insertion into skin. *Comput. Methods Biomech. Biomed. Eng.* 14(9):827–835, 2011.
5. N.T. Minh, H.L. Thanh, F. Karlsen, Analytical and numerical approaches for optimization design of MEMS based SU-8 microneedle, in *Biomedical informatics and technology*. Communications in Computers and Information Science, vol. 404, 2014. https://doi.org/10.1007/978-3-642-54121-6_9
6. S. Chakraborty, K. Tsuchiya, Development and fluidic simulation of microneedles for painless pathological interfacing with living systems. *J. Appl. Phys.* 103(11470):1–9, 2008.
7. D.B. das Olatinji, B. Al-Qallaf, Simulation based optimization of microneedle geometry to improve drug permeability in skin. *Proceeding of 7th* Industrial Simulation Conference, Loughborough, UK, 293–300, 2009.

8. F. Amin, S. Ahmad, Design, modeling and simulation of MEMS based silicon microneedles. *6th Vac. Surf. Sci. Conf. Asia Australia, J. Phys. Conf. Ser.* 439(012049):1–13, 2013.

9. www.microchem.com/pdf/SU-82000DataSheet2100and2150Ver5.pdf

10. B.P. Chaudhri, F. Ceyssens, P.D. Moor, C.V. Hoof, R. Puers, A high aspect ratio SU-8 fabrication technique for hollow MNs for transdermal drug delivery and blood extraction. *J. Micromech. Microeng.* 20, 2010. DOI: 064006-064012P-C

11. B. Al-Qallaf, D.B. Das, Optimization of square microneedle arrays for increasing drug permeability in skin. *Chem. Eng. Sci.* 63(9):2523–2535, 2008.

12. J.A. Bouwstra, M. Ponec, The skin barrier in healthy and diseased state. *Biochim. Biophys. Acta.* 1758(12):2080–2095, 2006.

13. M.S. Lhernould, Optimizing hollow microneedle arrays aimed at transdermal drug delivery. *Microsyst. Technol.* 19:1–8, 2013.

14. P. Sheshu, Textbook of finite element analysis. PHI Learning, 2012.

15. D.V. Hutton, *Fundamentals of finite element analysis*, 1st edition, Tata McGraw Hill, 2004. ISBN-0-07-112231-1

16. https://llis.nasa.gov/lesson/819

17. www.jntua.ac.in/gate-online-classes/registration/downloads/material/a158971425598.pdf

18. https://doc.comsol.com/5.5/doc/com.comsol.help.comsol/COMSOL_ReferenceManual.pdf

6 Fabrication and Characterization of Hollow Polymer Microneedles

6.1 INTRODUCTION

The pain and needle phobia associated with hypodermic needles have paved the way for discovery of newer technologies for drug administration to the human body through the skin. Hollow microneedle arrays offer versatility and control to the transdermal drug delivery systems where a variety of drugs and their continuous supply is concerned. They may be used as standalone devices or may be integrated with the drug reservoir or micropump. The fabrication of the hollow microneedles requires miniaturized structure fabrication. Over the years, the semiconductor industry has matured the microfabrication techniques which are now used with new materials and polymers for various applications. In the last chapter, we have seen the theoretical and structural analysis of hollow microneedles of different geometries. In this chapter we shall study in detail the fabrication steps involved in fabrication of hollow microneedles. One may choose and plan the fabrication technique depending upon the facilities available. Often, as in case of any other medical device, the choice of fabrication technique is mainly governed by the following factors:

(a) High and precise yield
(b) Fast process
(c) Assembly line compatibility and good control, and
(d) Economics

Out of the many structure types discussed for microneedles, we shall consider the fabrication of hollow cylindrical microneedles (example 5.1 and example 5.2) in this chapter elaborately. We have seen in the last chapter that a simple cylindrical hollow microneedle with SU-8 as material is able to pierce skin successfully without breaking. For making thick SU-8 microstructures, initially ultraviolet (UV) photolithography has been discussed. Subsequently, the drawback of UV photolithography in fabrication of such structures is discussed. Further, as a solution, maskless laser writing fabrication technique is discussed as a precise way for fabricating hollow SU-8 microneedles. The fabrication steps mentioned in this chapter lead to fabrication of a hollow microneedle array on a substrate with microfluidic conduits, which can be either attached to a syringe barrel or a micropump for pressurized fluid flow.

    DOI: 10.1201/9781003202264-6

6.2 INITIAL DEVICE SPECIFICATION

The application requirement, the theoretical and simulation-based analysis, available material and fabrication choices, form the basis of microstructure fabrication. As an example, we shall consider the fabrication of a hollow cylindrical microneedle array on a substrate whose specifications we have arrived at after many considerations and analysis as seen in earlier chapters. The substrate shall have through holes in it so that fluid can flow from the drug chamber to the microneedle lumen. The proposed schematic of the microneedle array and the substrate is shown in Figure 6.1. The device specifications are given in Table 6.1.

The specification of the microneedles has been proposed in earlier chapters. The rationale behind the specifications is revisited here.

(a) The needle array spacing is kept 500 µm to optimize between the reduction of the "bed of nails" effect and keeping the device size small [1].

(b) The dimensions of the flow channels are chosen in such a way to match the microneedle lumen opening (40 µm).

(c) The specifications for the flow channels in substrate have also been selected to match the fabrication aspects. These flow channels are to be fabricated in silicon. One of the preferred methods could be wet etching of silicon p-type wafer using potassium hydroxide (KOH). Due to crystallographic plane of silicon, KOH etching results in flow channels having [1 1 1] walls at 54.74° from the [1 0 0] plane. Hence to have a flow channel in a 275 µm silicon wafer, one side opening of 400 µm square is defined and on the microneedle side an opening of 65 µm is chosen. These grooves meet each other at a width of 40 µm (Figure 6.1c).

The next section shall describe the fabrication steps of hollow microneedles in detail.

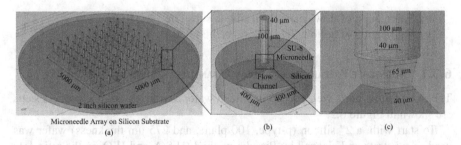

FIGURE 6.1 (a) Proposed schematic of 10X10 SU-8 hollow cylindrical microneedle array on 2 inch silicon wafer having flow channels, (b) magnified view of a single microneedle on silicon substrate and the flow channel and (c) magnified view of the flow channel and microneedle intersection point with dimensions.

TABLE 6.1

Proposed Specification of the Microneedle

S/No.	Parameter	Value
	Microneedle specifications	
1	Shape	Hollow cylindrical
2	Material	SU-8
3	Length	500 µm
4	Outer diameter	100 µm
5	Inner diameter	40 µm
6	Array size	10X10
7	Microneedle centre-to-centre spacing	500 µm
	Substrate specifications	
8	Material	Silicon wafer
9	Wafer thickness	275 µm
10	Wafer diameter	2 inch
11	Microneedle side opening	40–70 µm
12	Drug chamber side opening	400 µm

6.3 FABRICATION OF HOLLOW MICRONEEDLES

We shall now discuss one of the ways in which the device proposed in section 6.2 may be fabricated. It is important to understand that there are a lot of approaches in which this structure could be fabricated. One of the approaches is presented here so that the readers can understand holistically and appreciate the intricate fabrication steps that are involved in fabrication of such minute structures. It presents the approach which was adopted by Mishra et al. 2018 [2] with regard to the laboratory and cleanroom facilities available, the scalability and price of the end product involved. With time and technology, these steps shall indeed be replaced by better approaches and technology.

Following steps are adopted for hollow microneedle array fabrication.

(a) Fabrication of flow channels in silicon substrate
(b) Fabrication of hollow SU-8 microneedle

6.3.1 SUBSTRATE PREPARATION – FLOW CHANNELS IN SILICON

The microfluidic ports were etched in silicon wafer as a first step. The process steps are shown in Figure 6.2.

To start with, a 2" silicon (p-type, 100 plane, and 275 µm thickness) wafer was used as substrate and cleaned by Piranha method (H_2SO_4 and H_2O_2 in the ratio 1:1) for 30 minutes and blown dry with nitrogen. Figure 6.3 shows the process steps used for fabrication of hollow SU-8 microneedles. Dehydration bake was carried out in the oven for 30 minutes to drive away the excess moisture from the wafer. the 0.8 µm thick SiO_2 layer was formed on the silicon by dry-wet-dry oxidation technique in the Tempress Oxidation furnace (Netherlands) as a mask layer for wet etching

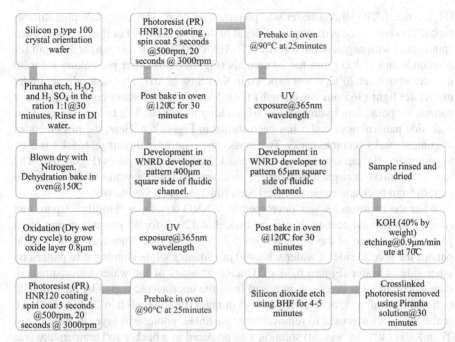

FIGURE 6.2 Flowchart for the steps involved in oxidation, both side patterning by UV photolithography and wet etching of the fluidic channels in a silicon wafer.

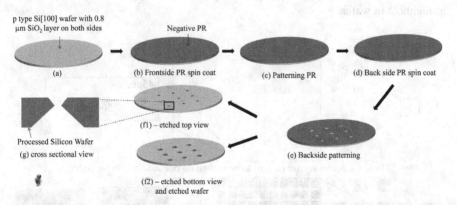

FIGURE 6.3 Process steps for fabrication of flow channels in silicon (a) oxidized silicon (p-type 100) substrate, (b) Spinning of negative photoresist on the wafer followed by pre-bake, (c) patterning of photoresist by photolithography to form 400 µm square followed by post bake, development, rinsing and drying, (d) spin coat of photoresist on second side of wafer following similar process as the other side, (e) patterning of second side of wafer by photolithography followed by post baking, development and drying followed by BHF etching of silicon dioxide and photoresist removal, (f1, f2) flow channels fabricated by KOH etching of silicon from both side patterned square and (g) cross-sectional view of the KOH etched silicon wafer having flow channels.

(Figure 6.3a). The silicon wafer was patterned on both sides using UV photolithography. This technique is discussed in detail in the next section. For this, the wafer was spin coated with negative photoresist (PR) HNR120 (Fujifilm, Japan) at 500 rpm for 5 seconds and at 3000 rpm for 20 seconds (Figure 6.3b). After pre-baking it for 30 minutes in oven at 90°C, it was exposed in Karl Suss MA6 (Italy) mask aligner under ultraviolet light (365 nm wavelength) for 6.5 seconds for pattern containing square pattern for potassium hydroxide (KOH) etching of silicon. The two masks used for both side patterning of wafer has been shown in Figure 6.4. Here, the placement of alignment marks becomes tricky. Two masks are required in this stage. Mask 1 is used to pattern the 400 μm size squares while mask 2 is for 65 μm squares (Figure 6.4). The top surface mask is aligned with pattern on the rear side of the wafer by observing the pattern from backside camera of Karl Suss Mask Aligner, Germany.

After the exposure, it was developed in WNRD developer (Fujifilm, Japan) and rinsed with n-butyl-acetate. It was postbaked at 120°C for 30 minutes giving resist pattern on one side of the wafer (Figure 6.3c). Similar procedure was carried out to pattern the second side of wafer with 400 μm squares while aligning it to pattern on other side of wafer (Figure 6.3d–e). Once both sides of the wafer were patterned, then silicon dioxide layer was removed from the unprotected region by etching with buffered Hydrofluoric acid (BHF) for 4–5 minutes. Piranha solution (H_2SO_4 and H_2O_2 in the ratio 1:1) was used to remove the cross-linked photoresist layer in 30 minutes. Then KOH (40% by weight) solution was prepared in a beaker and temperature was maintained at 70°C. Through holes were etched in silicon with the etch rate of 0.9 μm per minute. The sample was rinsed and dried. Figure 6.3 f1 and f2 show both the sides of the etched silicon wafer while Figure 6.3g shows the magnified schematic of the throughhole in wafer.

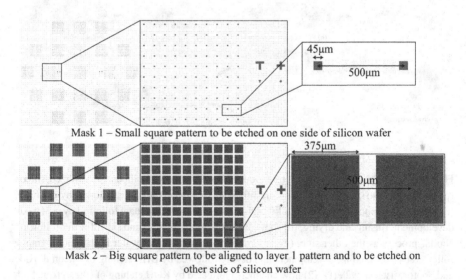

Mask 1 – Small square pattern to be etched on one side of silicon wafer

Mask 2 – Big square pattern to be aligned to layer 1 pattern and to be etched on other side of silicon wafer

FIGURE 6.4 Masks used for both side patterning of silicon wafer to yield etched flow channels in substrate.

6.3.2 HOLLOW SU-8 MICRONEEDLE ARRAY FABRICATION BY UV PHOTOLITHOGRAPHY

UV lithography is a technique to transfer patterns on a substrate by etching. Kang et al. (2006) [3] showed that the total light intensity distribution is affected by Fresnel diffraction, attenuation and reflection from substrate surface (Figure 6.5a).

Analytically, it can be represented as given below

$$I(x,z) \cong \left\{ \alpha'(z)\frac{I_i}{2} + \alpha'(2z_1 - z)R\alpha'\frac{I_i}{2} \right\} \times$$
$$\left\{ \left[C(U_2) - C(U_1) \right]^2 + \left[S(U_2) - S(U_1) \right]^2 \right\} \tag{6.1}$$

Where I_i Intensity of incident UV light, z is apparent distance from the top surface of the photoresist to the wafer surface, z_1 is thickness of photoresist, α' is the attenuation coefficient of photoresist and C(U) and S(U) are Fresnel integrals written as

$$C(U) = \int_0^u \cos\left(\frac{\pi}{2}\omega^2\right) d\omega, S(U) = \int_0^u \sin\left(\frac{\pi}{2}\omega^2\right) d\omega \tag{6.2}$$

Where U (U_1 and U_2) is the Fresnel number written as

$$U_1 = x_1 \sqrt{\frac{2}{\lambda(z_2 + z)}}, U_2 = x_2 \sqrt{\frac{2}{\lambda(z_2 + z)}} \tag{6.3}$$

Here x_1 is the horizontal distance from mask slit centre to left edge of mask, x_2 is the horizontal distance from mask slit centre to right edge of mask, λ is the wavelength of UV light and z_2 is the air gap between mask and top surface of the photoresist.

Based on the intensity distribution shown in Figure 6.5a, the SU-8 profile tends to get tapered i.e. it shows negatively sloped sidewall. If the exposure dose is increased to correct the tapering profile, then slightly straight sidewalls are obtained, but at the cost of enlarged structure width. The negative sidewall can be somewhat corrected by employing methods like increasing the exposure time, reducing the gap between SU-8 film and mask or using glycerol to cover the air gap between the mask and the SU-8 film [4,5].

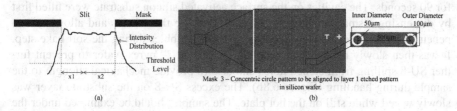

FIGURE 6.5   (a) Intensity distribution in UVL process and (b) darkfield outlay to be used as mask 3.

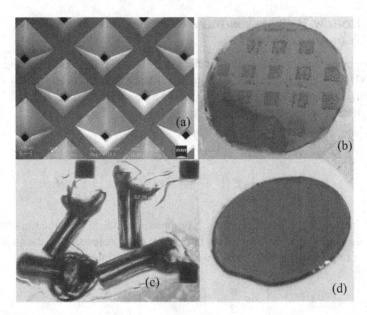

FIGURE 6.6   (a) SEM image of the KOH etched through holes in silicon wafer, (b) development of numerous air bubbles when SU-8 is spun on substrate containing through holes, (c) needles formed on the bubbles breaking away during development, and (d) SU-8 layer surface after optimization step of reducing the bubbles.

Keeping these limitations in mind, UV photolithography shall be used for fabrication of hollow microneedles on top of the flow channels etched in silicon. Once the substrate with flow channels is ready as discussed in the previous section, a pretreatment step is required to cover the grooves (flow channels) with SU-8 in order to present a uniform surface for final SU-8-layer coat on it. If the final SU-8 layer (500 μm) is spin coated directly on the silicon wafer having flow channels in it, then a lot of air bubbles would be generated just above the etched grooves in the substrate (Figure 6.6b). These bubbles pose a lot of problems during fabrication of microneedles where microneedles formed on top of this bubble agglomeration break easily during the development process (Figure 6.6c). Hence a pretreatment step is required to cover the grooves in silicon. Based upon the pretreatment step, the process steps are shown as flowchart in Figure 6.7 and important steps illustrated in Figure 6.8.

The wafer was treated with oxygen plasma (Zepto, Diener Electronic, Germany) for 90 seconds. The cavities on the surface activated silicon substrate were filled first by dispensing a small quantity of SU-8 2150 above these holes and allowing it to penetrate in the holes by slowly releasing the air bubbles during the soft bake step. It was then slowly manually aligned and placed on 2" pyrex wafer to prevent further SU-8 spilling from etched holes and also to provide mechanical strength to the sample during handling (Figure 6.8b). The excess SU-8 on the substrate layer was slowly wiped while still on the hot plate. The sample should be examined under the microscope from the transparent pyrex wafer side to ensure that no air bubble remains (Figure 6.6d). The sample was cooled and then the final SU-8 layer was spun.

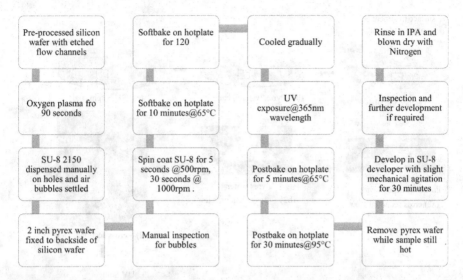

FIGURE 6.7 Flowchart for the process steps involved in hollow SU-8 microneedle array fabrication by UV photolithography.

Once the grooves in the silicon wafers were filled during the pretreatment step, then SU-8 2150 (MicroChem, USA) was spun first at 500 rpm for 10 seconds to evenly spread the viscous photoresist and then at 1000 rpm for 30 seconds to get the desired thickness. The samples were then softbaked on a hotplate at 65°C for 10 minutes and 95°C for 120 minutes. They were cooled down gradually to room temperature to avoid any cracks due to thermal stress (Figure 6.8c). Then, they were exposed with Karl Suss MA-6 Mask Aligner under UV light (365 nm) at varying light intensities. The intensity is varied by changing the exposure time of mask aligner as shown in Table 6.2 below.

The exposure step opens up the SU-8 ring which enhances the crosslinking efficiency. Post exposure bake was carried out for 5 minutes at 65°C and for 30 minutes at 95°C on hotplate (Figure 6.8d). This post exposure baking allows the photogenerated acid to diffuse in the photoresist and catalyze the reaction. After postbake step, the pyrex wafer is removed while SU-8 was still hot and in molten state as the SU-8 leaked in through holes quickly solidifies on cooling after the post exposure bake step and it becomes difficult to remove the pyrex wafer. Then the sample was cooled down gradually. It was developed in the SU-8 developer solution (MicroChem, USA) (Figure 6.8e). The uncrosslinked SU-8 which leaks in the grooves gets removed during the development stage. Once the sample is put in the developer solution, at an elevated platform, the developer removes uncrosslinked SU-8 from both sides of sample. It reduces the development time otherwise required for only single side of wafer available for the developer to act. Slight mechanical agitation by hand at this stage enables used developer to be removed and fresh developer to enter in the microneedle lumen area, thus resulting in complete uncross-linked SU-8 removal from such microsized space (40 μm). After development, the sample was rinsed in isopropyl alcohol and then slowly blown dry with nitrogen (Figure 6.8f).

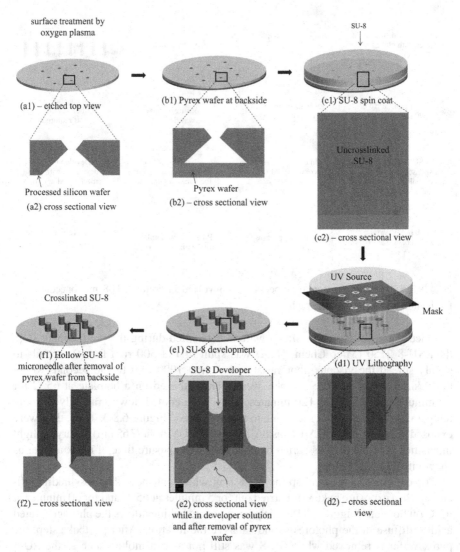

FIGURE 6.8 Process steps involved in hollow SU-8 microneedle array fabrication by UV photolithography.

6.3.3 CHARACTERIZATION OF MICRONEEDLES FABRICATED BY UV PHOTOLITHOGRAPHY

6.3.3.1 SEM and Optical Image-based Characterization

It is well known that SU-8 consists of eight epoxide groups in each monomer which is highly functional. During exposure to light or photons, the photoacid generator is absorbed by chemical reaction and produces an acid. It diffuses to SU-8 monomers and acts as a catalyst in the exposed regions by helping in cross linking. During

TABLE 6.2
List of Exposure Dose for SU-8 Microneedle Fabrication

S/No.	Exposure Time (s)	Intensity mJ/cm^2	S/No.	Exposure Time (s)	Intensity mJ/cm^2
1	50	740	6	600	8880
2	100	1480	7	700	10360
3	300	4440	8	800	11840
4	400	5920	9	900	13320
5	500	7400			

cross linking, the SU-8 epoxy groups are zipped together to form a network [6]. This highly crosslinked epoxy structure has high structural stability. The reaction rate of the crosslinking depends on the diffusion of the catalyst which in turn is dependent on the laser intensity. The post exposure bake (PEB) step promotes diffusion of the catalyst and mobility of SU-8 monomers, increasing the efficacy of the crosslinking process. Fast polymerization is required so that the catalyst does not further diffuse to the non-exposed areas. This cage effect is required for high resolution SU-8 structures [7]. The shape of the crosslinked structure mainly follows the irradiation pattern [3]. Scanning Electron microscopy was used to determine the top profile of the developed microstructures. Figure 6.9 shows SEM profiles of microstructures fabricated by UV photolithography [8]. In UV lithography, the exposure intensity can be varied only by increasing the exposure time which poses many limitations like long times during UV dosage optimization [5].

6.3.3.1.1 Top Profile Studies for SU-8 Microneedles

Scanning Electron microscopy was used to determine the top profile of the developed microstructures. Figure 6.9a–d shows SEM profiles of microstructures fabricated by UV photlithography. In UV photlithography, the UV dose can be varied by increasing the exposure time of the sample. For the dose of 4440 mJ/cm^2, a T shaped profile resulted since much of the incident UV energy is absorbed on the top of the resist layer giving features around 200 μm instead of 100 μm (Figure 6.9a). As the UV exposure dose is increased, the T-shape profile is corrected (Figure 6.9b). Best profile structures are obtained for 7400 mJ/cm^2 (Figure 6.9c) which is 500 s of UV exposure. But the increased dose resulted in enlarged features with negative slope again (Figure 6.9d).

6.3.3.1.2 Cross-sectional Profile Studies for SU-8 Microneedles

To get a better picture and understanding of the fabricated structure, cross-sectional profile also needs to be studied. Figure 6.10 shows the cross-sectional profile of fabricated microneedles with different doses. Figure 6.10a shows the actual optical micrographs with boundary highlighted while Figure 6.10b shows the plot obtained by tracing one side of the cross-sectional profile of the fabricated microneedle at particular intensity. As discussed for the top profile views, in case of UV photolithography (Figure 6.10b), it is seen that due to diffraction the top surface of the resist is

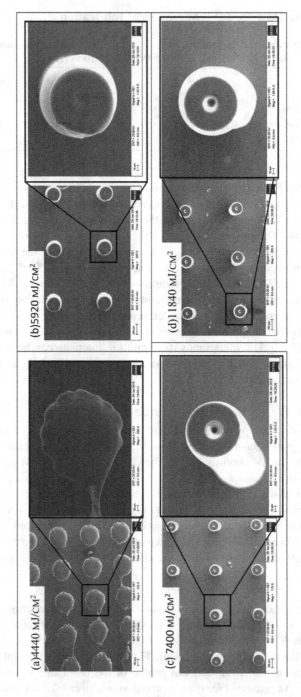

FIGURE 6.9 SEM micrograph of microneedles fabricated at different intensities by UVL. Comparing the top profiles, a dose of 7400 mJ/cm² is found to be optimum for UV photolithography.

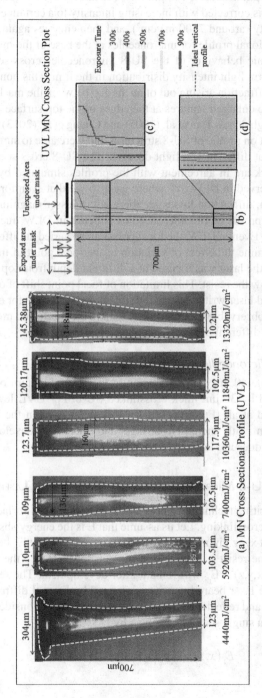

FIGURE 6.10 Optical microscopic image of cross-sectional profile of cylindrical microneedles fabricated by UVL. Plot of developed cross-sectional profile for microneedles with increasing dosage for UVL (c). The effect of different UV intensities on microneedle top (d) and bottom (e) microneedle structure are shown in inset.

enlarged around 300 μm instead of 100 μm for 4440 mJ/cm^2 dose (Figure 6.10c). This feature enlargement is corrected with increasing intensity to a certain extent which is the optimum intensity (around 7400 mJ/cm^2), and then enlarges again. Figure 6.10d shows the cross-sectional profile for microneedles at the base of the microneedle.

The non-monotonic behaviour of the SU-8 microneedles cross sections results due to the non-uniform light intensity distribution in the film. This non-uniformity is caused by Fresnel diffraction arising out of an air gap between the mask plate and the sample. This leads to enlarged features at the edges of the top surface of the resist at very low and very high doses. Tian at al. (2005) and Chung et al. (2013) [9,10] studied this diffraction effect on profile of SU-8 structures and were able to simulate different structures resulting at different UV light dosages. The SU-8 cross-sectional profiles obtained in this work are in agreement with the profiles simulated by these works. Same pattern is observed in DLW too, where we also see that even for laser induced photopolymerization, diffraction losses are there but of lesser magnitude than UVL. The cross-sectional profile of the microneedles is better in DLW due to the highly coherent laser beam. Even in DLW, the focal point can be varied at different thickness of the resist. To examine the effect of variation of the focal point of the laser within the resist thickness, the laser was focused at a different depth from top surface of the resist to 100 μm below the resist. This limitation of focal point shift of only 100 μm in the resist for the used laser writer system did not give much scope for examining this effect. No optically observable difference was seen in the fabricated microneedles by focusing the laser at different depths within the resist.

6.3.3.1.3   SU-8 Microneedle Lumen Studies

Next, we consider the 40 μm diameter lumen in the desired pattern of microneedle which is surrounded by 30 μm thick crosslinked SU-8. For UVL lower doses, the lumen is not defined at all (Figure 6.11a–b). It is defined when the UV dosage is increased. But even with increased dosage and increased development times, uncrosslinked SU-8 does not leave the lumen space.

6.3.4   Hollow SU-8 Microneedle Fabrication by Direct Laser Writing

In the direct laser writing technique, a laser beam incident on the surface of the resist provides energy for crosslinking. Let us assume that E is the energy absorbed per unit volume around point x,y,z giving spatial distribution E(x,y,z). Let E_{th} be the threshold dose above which the resist is insoluble in the developer. Hence the parts of SU-8 receiving energy E(x,y,z) ≥ E_{th} will remain after development. The resist surface is denoted by z=0. The laser beam intensity is affected by Fresnel diffraction, absorbance in the polymer, and reflectance from the substrate surface. Considering these, the energy provided by a single scan line is given by [133]

$$E_1\left(x,z\right) = E_{line}\left(x,z\right) + rE_{line}\left(x,z'\right) \qquad (6.4)$$

Where, $z' = 2z_1 - z$ and $0 " z" z_1$. E_{line} is given by

$$E_{line}(x,z) = \frac{\sqrt{2}}{\pi} \frac{\alpha \kappa P}{vw(z)} \exp\left[-\left(2\frac{x^2}{w(z)} + \alpha z\right)\right] \quad (6.5)$$

Where E_{line} is the energy distribution for single scan line, r is the reflection factor, P is the laser power of the system and κ is the loss factor considering losses due to mirrors and lenses in beam path and reflection loss at resist surface, w(z) is the beam waist and is given by

$$w(z) = Mw_0 \left[1 + \left(\frac{z}{z_R}\right)^2\right]^{\frac{1}{2}} \quad (6.6)$$

Where w_0 is the spot radius at z=0 and $z_R = \pi\left(w_0^2/\lambda\right)$ is the Rayleigh length. An SU-8 structure is generated by scanning N laser lines separated by a distance d_x. Hence the total energy distribution for a structure is given by adding the energy distributions for each line. The threshold energy is fixed for a system. Hence the resulting SU-8 structure profile can be changed by changing the laser power. E Rabe et al. (2007) [11] showed that for low powers, the structure width is smaller and is somewhat undercut. When the laser power is increased, the profile obtains a negative slope. If the power is increased further, large width structures result as is the case with the over exposed samples in UV lithography. With these considerations, SU-8 microneedles with 100 μm outer diameter, 40 μm inner diameter and height 500 μm are targeted to be fabricated.

The process for direct laser writing remains the same as UV photolithography except for the exposure. Instead of using a mask, pattern (Figure 6.11) is written directly on the SU-8 coated substrate. The process flow is shown in Figure 6.12. The substrate was exposed with a continuous laser beam (Helium–Cadmium laser at 325 nm, Microtech LW-405B, Italy, laser spot size 5 μm) at different intensities (Table 6.3) to optimize the appropriate SU-8 dosage. The alignment of the hollow

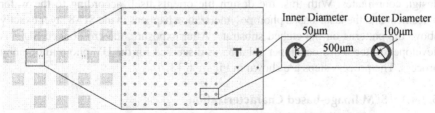

Design pattern 3 – Concentric circle pattern to be aligned to layer 1 etched pattern in silicon wafer and to be directly written by laser on the SU-8 layer

FIGURE 6.11 Pattern to be written on SU-8 coated substrate by DLW process.

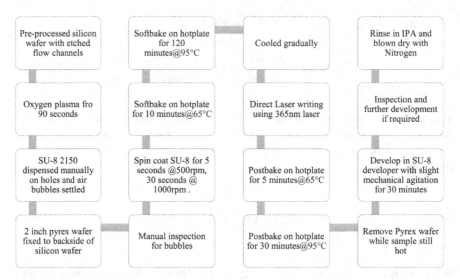

FIGURE 6.12 Flowchart for the process steps involved in hollow SU-8 microneedle array fabrication by direct laser writing.

TABLE 6.3
List of Different Exposure Intensity of Direct Laser Writing Technique

Intensity mJ/cm²	Intensity mJ/cm²
901	1774
1185	1916
1277	
1426	
1639	

MNs was done on the through holes on substrate using alignment technique of the laser writer by defining two points (A&B) in the pattern on the wafer with known design coordinates. With this, the design file orients itself according to the wafer placement and the laser-based photopolymerization happens in the SU-8 layer exactly above the patterns on the etched substrate. After exposure, the crosslinked SU-8 is developed in the SU-8 developer solution as discussed for the UV photolithography process. The process steps are shown in Figure 6.13.

6.3.4.1 SEM Image-based Characterization

In direct laser writing technique, a laser beam incident on the surface of the resist provides energy for crosslinking. Laser induced photopolymerization occurs in SU-8 around the beam waist where the intensity of the optical field is highest and exceeds

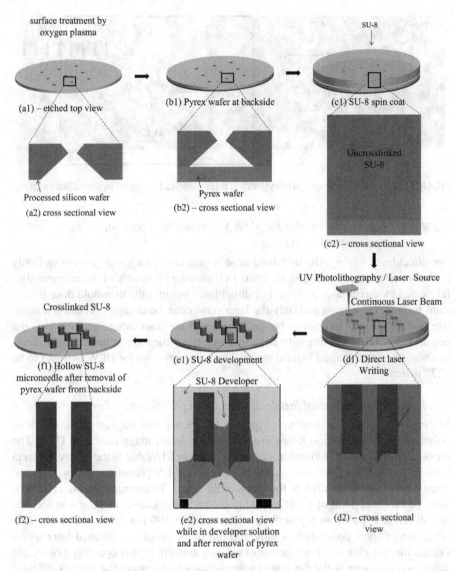

FIGURE 6.13   Process steps for hollow microneedle fabrication by direct laser writing.

the threshold polymerization energy of SU-8 [11]. The threshold energy is fixed for a system.

Figure 6.14 shows the SEM image of the microneedle fabricated at optimum direct laser writing doses. In case of DLW, hollow lumen is developed and can be seen in the SEM image. The tightly focused Helium–Cadmium laser beam (365 nm) used for direct laser writing has lesser diffraction related losses than UV photolithography technique. It also avoids the problem of undesired SU-8 cross linking. While designing, the microneedle wall should be less than 30 µm thickness to accommodate the SU-8 feature broadening.

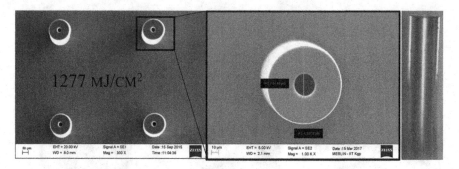

FIGURE 6.14   SEM micrograph of cylindrical microneedles fabricated by direct laser writing.

6.3.4.1.1   Top Profile Studies for SU-8 Microneedles Fabricated by Direct Laser Writing

For direct laser writing, the threshold dose is constant for a given process and only the laser power can be changed. Figure 6.14 shows the top profile of the microneedles fabricated by direct laser writing. For direct laser writing, the threshold dose is constant for a given process and only the laser power can be changed. There is a negative profile at lower intensities but on increase of the laser intensity, it corrects the negative profile. If the intensity is further increased then the structure again tends to undercut with enlarged feature width. The optimum dose for DLW is found to be 1277 mJ/cm^2.

6.3.4.1.2   Cross-sectional Profile Studies for SU-8 Microneedles

In direct laser writing, as seen in Figure 6.13, we see that for laser induced photo polymerization, diffraction losses are there but of lesser magnitude than UVL. The cross-sectional profile of the microneedles is better in DLW due to the highly coherent laser beam. In direct laser writing, as was the case for UV photolithography, the focal point can be varied at different thickness of the resist. To examine the effect of variation of the focal point of the laser within the resist thickness, the laser was focused at a different depth from top surface of the resist to 100 μm below the resist. This limitation of focal point shift of only 100 μm in the resist for the used laser writer system did not give much scope for examining this effect. No optically observable difference was seen in the fabricated microneedles by focusing the laser at different depths within the resist.

6.3.4.1.3   SU-8 Microneedle Lumen Studies

When direct laser writing technique is used, lumen could be clearly seen in SEM image (Figure 6.13) and hollow lumen can be traced in the corresponding cross-sectional image which has extended almost till the substrate. This development of hollow lumen is very important if the microneedles are intended for drug delivery applications. The spatial coherence of the laser beam and the large power concentrated in the beam provides great directivity to induce cross linking reactions in thick films.

In case of direct laser writing, hollow lumen is also developed and can be seen in the SEM image. The tightly focused Helium–Cadmium laser beam (365 nm) used for direct laser writing has lesser diffraction related losses than UV photolithography technique. It also avoids the problem of undesired SU-8 cross linking. While designing, the microneedle wall should be less than 30 μm thick to accommodate the SU-8 feature broadening. In our case, we have chosen outer and inner diameter as 95 μm and 55 μm to yield desired microneedle geometry of inner diameter 40 μm and outer diameter of 100 μm.

6.3.4.2 Alignment of Microneedles on Flow Channels

It is equally important that the SU-8 microneedles are perfectly aligned on the through holes in silicon wafer. Figure 6.15a–b shows the SEM images during the alignment optimization steps where the microneedles are misaligned until finally optimized to cover the through hole entirely (Figure 6.15c). In order to get a better understanding of the developed microneedles, a cross-sectional SEM image of the microneedles on the etched flow ports was obtained (Figure 6.15d–e). The roughness visible in the silicon wafer cross section is due to uneven breaking of wafer by tweezers.

6.4 MECHANICAL CHARACTERIZATION OF MICRONEEDLES

6.4.1 NANOINDENTATION-BASED MECHANICAL CHARACTERIZATIONS

Nanoindentation testing is a fairly mature technique which uses the recorded depth of penetration of an indenter into the specimen along with the measured applied load to determine the area of contact and hence the hardness of the test specimen [12]. Many other mechanical properties can also be obtained from the experimental load–displacement curve, the most straight-forward being the elastic modulus. The Hysitron triboindentor was used to characterize the mechanical properties of the microneedles. A Berkovich shaped indenter was used and Oliver Pharr method was used to calculate the hardness and Young's modulus of the cross linked SU-8 as shown in Figure 6.16. The hardness is given by

$$H = \frac{P_{max}}{A_c} \tag{6.7}$$

Where P_{max} is the maximum load and A_c is the indentation area. The contact area is calculated based on the indenter tip shape and penetration contact depth. The measured total penetration depth, δ_T, includes both the plastic (contact) depth, δ_c, and the elastic depth, δ_e. For an ideal Ber kovich indenter $A_c = 24.5\delta^2_c$.

Young's modulus can be determined from the slope of the unloading curve using a modified form of Sneddon's flat punch equation where

$$S_u = \gamma\beta\frac{2}{\sqrt{\pi}}E_r\sqrt{A_c} \tag{6.8}$$

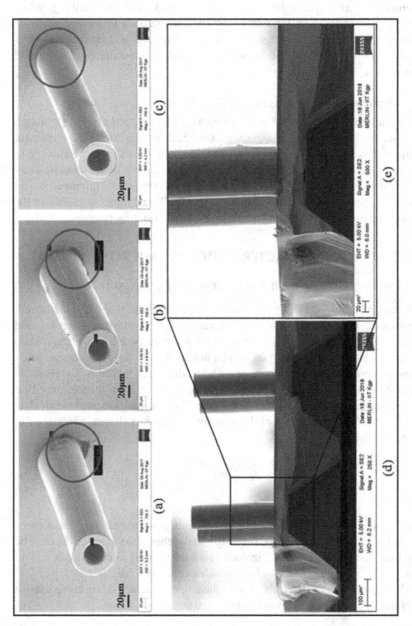

FIGURE 6.15 SEM images showing (a) the misaligned SU-8 microneedle on the pre-etched silicon wafer, (b) microneedle alignment corrected to some extent and (c) SEM image of well aligned microneedle, (d) cross-sectional view of the hollow microneedles on the etched flow channel in silicon wafer and (e) zoomed view of the microneedle cross section showing proper alignment of microneedle.

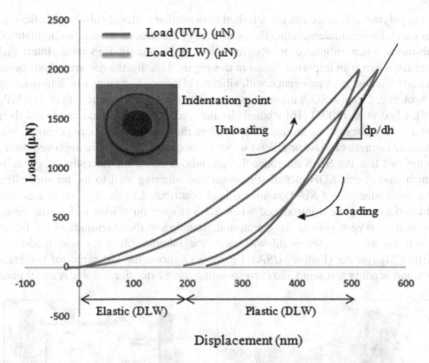

FIGURE 6.16 Load–displacement curve for hollow SU-8 microneedles (500 μm length, 100 μm outer diameter and 40 μm inner diameter fabricated by UVL and DLW.

where S_u is slope of the unloading curve, E_r is the elastic reduced modulus which can be derived from Young's modulus (E) and Poisson's ratio(v) of the indenter and the test material via

$$\frac{1}{E_r} = \frac{1 - v_m^2}{E_m} + \frac{1 - v_i^2}{E_i} \tag{6.9}$$

where the subscripts m and i refer to the test material and indenter, respectively.

Figure 6.16 shows the load–displacement data for the indenter, as it was driven into and pulled out into the SU-8 microstructures. Indentation performed on SU-8 microneedle wall indicates the moderate viscoelastic behavior with a hysteresis loop between the loading and unloading curves.

6.4.2 FORCE–DISPLACEMENT-BASED MECHANICAL CHARACTERIZATION

In order to prove that the hollow microneedles can be used as biological interface for transdermal drug delivery, their maximum bending and compression forces should be higher than the skin resistive forces so that they do not break during insertion or retraction. In our previous work, we have calculated the resistive forces offered by skin during skin insertion for microneedles of the chosen geometry [13]. In our work, polymer microneedles are attached to dissimilar material substrates (silicon in our case). Hence, understanding the behaviour of the structures on different material substrates when subjected to progressive load is equally important. Interfacial strength is hence an important factor in this regard. Usually, the negative photoresists like SU-8 show good adherence with silicon [14]. For SU-8, it can be improved by heat curing. Circular SU-8 microposts were found to have yield strength of 112 MPa [15]. Khoo et al. (2003) [16] studied circular microposts on silicon and found their shear failure stress to be around 7 Mpa before the interfacial fracture occurred. The adhesive strength of SU-8 on silicon wafer in our case is all the more improved owing to the fact that the SU-8 microneedles are fabricated on the microfluidic conduits which have rough KOH etched silicon surface adhering well to the microneedles. The good adhesion of SU-8 on silicon is also attributed to the high surface energy and surface roughness introduced when SU-8 is spun on silicon wafer after being exposed to oxygen plasma. To get an understanding of the magnitude of the forces which the microneedles could withstand, the microneedle array was loaded on Instron Microtester (Instron, USA). Figure 6.17 shows the test setup for compression and bending test setup and corresponding force extension graphs. A metal plate

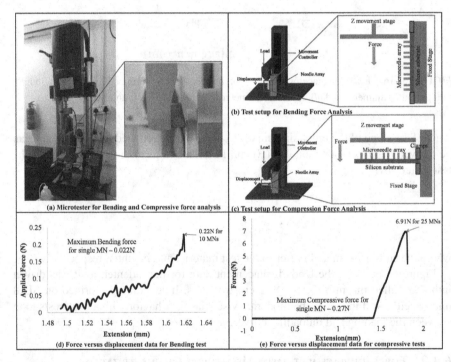

FIGURE 6.17 (a) The microtester equipment used for force–displacement-based mechanical characterizations of microneedles with zoomed view in inset, the test setup for (b) bending and (c) compression tests with the zoomed view of the moving z-stage and sample alignment in inset, (d) load versus extension curve for bending test data and (e) load versus extension curve for compression test data for microneedles (500 μm length, 100 μm outer diameter and 40 μm inner diameter) fabricated by direct laser writing.

was driven towards the microtubes until they broke. The needles will not break if the applied load is below the maximum compressive or bending force. As the z-stage applies increased load, eventually the needles break. The sharp drop in the curve (Figure 6.17d–e) marks the microneedle fracture point [17].

Maximum compressive force for breaking an array of 25 microneedles fabricated by direct laser writing is found to be 6.91 N, or 0.27 N per microneedle. Similarly, maximum bending force for 10 microneedles is 0.22 N or 0.022 N per microneedle. From the optical micrograph of the broken microneedles (Figure 6.17a) during bending test, it can be seen that the needles broke near their bottom due to the inter-facial fracture (i.e. delamination) rather than breaking. The body of the majority of microneedles show no failure. This may be owed to the high Young's modulus of SU-8 (4–5 GPa). Some microneedles which appear to be broken in the needle body may be due to the crack propagated by the voids. These voids are created due to the impur-ities in the SU-8 layer which are introduced during the fabrication stage. Hence the mechanical characterization tests establish that the microneedles are strong enough to penetrate skin successfully having their maximum compression and bending forces much higher than skin resistive forces.

6.5 MICROFLUIDIC TESTS

After the fabrication of SU-8 microneedles, it is important to ensure that the microneedles are completely hollow since their aim is to effectively transport the liquid drug formulation from drug reservoir to skin. It is also required to establish the micropump readiness of the microneedle system. Figure 6.18b–c shows the top and bottom view of the fabricated microneedles. To test whether the microneedles are completely hollow, the microneedle array is first observed under the microscope (transmission mode) (Figure 6.18d) as followed by Vinayakumar et al. (2014) [18]. The light shining from the microneedles is guided through the through holes and then through microneedle lumen. It confirms that the through holes are thoroughly developed. For microfluidic characterization, this microneedle array on silicon is attached to the 5 ml syringe opening and water is passed through the microneedle array at varying pressure drops at inlet. Figure 6.19a shows the test setup for flow rate characterization.

Deionized Water is passed through the microneedle array at various set pressures. The desired water pressure is achieved by pushing nitrogen through a deionized water filled chamber. Pressure is monitored by an Abus PT series transducer near the syringe inlet. Initially, for low pressure (<9 kPa), small water droplets come out of the microneedle array. On increasing the pressure, water jet is issued out of the microneedles (water jet at 8 kPa pressure is shown in Figure 6.18f). Figure 6.19 shows the variation of the experimentally determined flowrates with inlet pressure drops where flow rate increases with pressure. At 8 kPa pressure difference, flow rate of 1.7 μL/s per microneedle array and consequently 0.017 μL/s per microneedle is achieved.

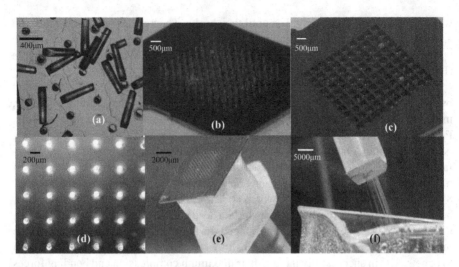

FIGURE 6.18 (a) microneedles after bending test showing mostly interfacial delamination instead of breaking in the microneedle body, top (b) and bottom (c) view of the fabricated microneedle array, (d) Light illumination test for SU-8 microneedles, (e) microneedle array mounted on the syringe for microfluidic tests, and (f) water jet coming out of microneedles when fixed in front of a 5 ml syringe and water pressure applied at 9 kPa showing that microneedles are hollow with 100% yield.

6.6 BIOLOGICAL SKIN INSERTION-BASED CHARACTERIZATION

Mechanical interaction characterization of the hollow microneedles with skin was performed in steps of three tests (Figure 6.20 and 6.21). Firstly, the microneedle array (dipped in methylene blue) was driven perpendicularly against freshly excised mouse skin (6–8 weeks old Swiss Albino) using a customized platform (Test 1) (Figure 6.20a–d). This was done by fixing the microneedle array on the stage which can be moved manually to be pressed against the mice skin kept on the platform as shown in Figure 6.20b. The microneedle array was brought in contact with mouse skin, pressed 0.5 mm further and then slowly withdrawn. Second, the microneedle array was pressed against mice (Swiss Albino) in the abdomen area (test 2) (Figure 6.21a–c). Then finally as test 3, the microneedle array was similarly pressed against lab rat (Rattus Norvegicus) skin in the abdomen area (Figure 6.21d–f). The protocol was approved under the ethical clearance from the Committee for the Purpose of Control and Supervision of Experiments on Animals (CPCSEA), New Delhi, India. On visual examination, in all the three tests, the mouse and the rat skins showed the marks of all 10X10 microneedles, showing absence of "bed of nails" effect with 500 μm array spacing [19]. The microneedle array was examined to see that all the microneedles remained intact after inserting the microneedle array after 10 insertions at different places of mice skin. Figure 6.20d shows the SU-8 microneedles to be intact even after multiple insertions. Hence, fabricated SU-8 microneedles with 500 μm array spacing successfully penetrated the mice and rat skin. These skin models are widely

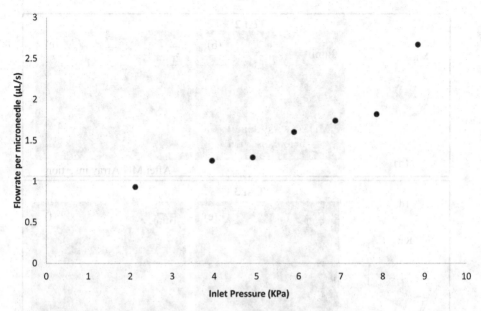

FIGURE 6.19 (a) Test setup for flowrate measurement and (b) Flowrate per microneedle at different inlet pressure difference.

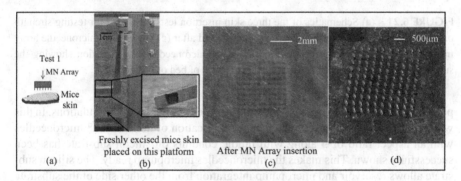

FIGURE 6.20 (a) Schematics of the skin insertion tests, (b) Test 1 setup employed for pressing microneedle array in excised mouse skin, (c) insertion marks of MNs on mouse's skin and (d) intact microneedles after multiple skin insertions.

acceptable as human skin models, but further studies on skin similar to human skin type is required.

6.7 CONCLUSION

Hollow microneedle fabrication as a transdermal drug delivery device has not seen much commercial light due to the complex fabrication steps involved. Yet, it holds the

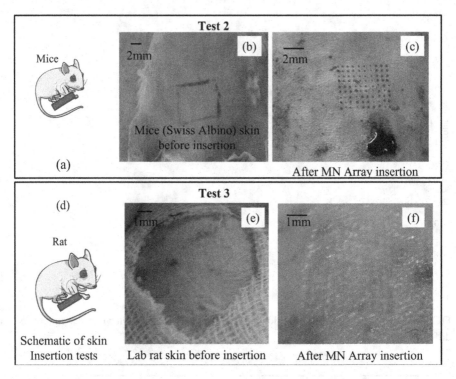

FIGURE 6.21 (a) Schematics of the three skin insertion tests employed for testing strength of microneedles. Images of mouse skin before (b) and after (c) pressing of microneedle array in test 2 (d), and rat skin (e) before and (f) after microneedle array insertion showing the successful microneedle array penetration and absence of bed of nails effect.

potential of being able to be used to handle all types of liquid drug formulations. In this work, A MEMS-based novel technique for fabrication of hollow SU-8 microneedles with an aspect ratio of 5 along with fluidic conduits on silicon substrate has been successfully shown. This makes the microneedles micropump ready. The silicon substrate allows reservoir and micropump integration from the other side of the substrate and also provides good mechanical support. The hollow SU-8 microneedles have been fabricated by DLW with fully developed needle lumen, allowing the scalable and batch production of such structures. The maximum compression and bending forces of the microneedles indicate that they shall pierce human skin successfully without any breakage. The microneedles penetrated successfully in mice skin without breaking. They have flow rate of around 0.017 µL/s per microneedle at 8 kPa pressure difference at the inlet. These microneedles being almost an order stronger than that required for human skin insertion, it is still desirable to increase their strength and hardness so that they can handle storage and handling stresses. Such a process is described in the next chapter.

REFERENCES

1. Z. Xiang, H. Wang, S.K. Murugappan, S.C. Yen, G. Pastorin, C. Lee, Dense vertical SU-8 MNs drawn from a heated mold with precisely controlled volume. *J. Micromech. Microeng.* 25:025013–025023, 2015.

2. R. Mishra, T. Maiti, T. Bhattacharyya, Development of SU-8 hollow microneedles on silicon substrate with microfluidic interconnects for transdermal drug delivery. *J. Micromech. Microeng.* 28:105017, 2018.

3. W.J. Kang, E. Rabe, S. Kopetz, A. Neyer, Novel exposure methods based on reflection and refraction effects in the field of SU-8 lithography. *J. Micromech. Microeng.* 16:821–31, 2006.

4. Y.J. Chuang, F.G. Tseng, K.W. Lin, Reduction of diffraction effect of UV exposure on SU-8 negative thick photoresist by airgap elimination. *Microsyst. Technol.* 8:308–13, 2002.

5. M.B. Zhang, S. Chan-Park, R. Connera, Effect of exposure dose on the replication fidelity and profile of very high aspect ratio microchannels in SU-8. *Lab Chip.* Dec; 4(6):646–53, 2004.

6. J.M. Shaw, J.D. Gelorme, N.C. LaBianca, W.E. Conley, S.J. Holmes, Negative photoresists for optical lithography. *IBM J. Res. Dev.* 41, 1997.

7. M. Despont, H. Lorenz, N. Fahoni, J. Brugger, P. Vettiger, P. Renaud. High aspect ratio, ultrathick negative-tone near-UV photoresist for MEMS application. *Proceedings of 10th IEEE International Workshop on Micro Electro Mechanical Systems (MEMS'97)*, Nagoya, Japan, 518–522, 1997.

8. R. Mishra, T.K. Bhattacharyya, T.K. Maiti, Structural Comparison of SU-8 Microtubes fabricated by Direct Laser Writing and UV Lithography, *IEEE Sensors Conference*, Glasgow, Scotland, 2017.

9. L. Tian, S.A. Greenberg, S.W. Kong, J. Altschuler, I.S. Kohane, P.J. Park, Discovering statistically significant pathways in expression profiling studies. *Proc. Natl. Acad. Sci. U S A.* Sep 20; 102(38):13544–13549, 2005. DOI: 10.1073/pnas.0506577102

10. S. Chung, S. Park, Effects of temperature on mechanical properties of SU-8 photoresist material. *J. Mech. Sci. Technol.* 27:2701–2707, 2013. https://doi.org/10.1007/s12206-013-0714-6

11. E. Rabe, S. Kopetz, A. Neyer, The generation of mould patterns for multimode optical waveguide components by direct laser writing of SU-8 at 364 nm. *J. Micromech. Microeng.* 17:1664–1670, 2007.

12. B. Bhushan (ed.), *Springer Handbook of Nanotechnology*, Springer.

13. R. Mishra, T.K. Bhattacharyya, T.K. Maiti, Theoretical analysis and simulation of SU-8 microneedles for effective skin penetration and drug delivery. *IEEE Sensors Conference*, Busan, South Korea, 2015.

14. M. Madou, *Fundamentals of Microfabrication*, Boca Raton, FL: CRC Press, 1997.

15. C. Ishiyama, M. Sone, Y. Higo, Effects of heat curing on adhesive strength between microsized SU-8 and Si substrate. *23rd European Mask and Lithography Conference (EMLC)*, Grenoble, France, 2007.

16. H.S. Khoo, K.K. Liu, F.G. Tseng, Mechanical strength and interfacial failure analysis of cantilevered SU-8 microposts. *J. Micromech. Microeng.* 13:822–831, 2003.

17. S.P. Davis, B.J. Landis, Z.H. Adams, M.G. Allen, M.R. Prausnitz, Insertion of microneedles into skin: Measurements and prediction of insertion force and needle fracture force. *J. Biomech.* 37:1155–1163, 2004.

18. K.B.Vinayakumar, G.M. Hegde, M.M. Nayak, N.S. Dinesh, K. Rajanna, Fabrication and characterization of gold coated hollow silicon microneedle array for drug delivery. *Microelectron. Eng.* 128:12–182014.
19. Z. Xiang, H. Wang, S. K. Murugappan, S.C. Yen, G. Pastorin, C. Lee, Dense vertical SU-8 MNs drawn from a heated mold with precisely controlled volume. *J. Micromech. Microeng.* 25:025013–025023, 2015.

7 C-MEMS Process-based Post Fabrication Strength Enhancement of Microneedles

7.1 INTRODUCTION

The search for new materials, dimensions and design for microneedles has gained momentum recently due to their increasingly accepted role in a variety of applications like vaccine delivery, cancer treatment and emergency drug management. This search is centered on desired material properties like biocompatibility, high strength to break through the stratum corneum (outermost layer of skin) and controlled delivery of drugs. The initial research on microneedles was directed towards silicon as it was widely accepted by the microelectronics and MEMS industry [1,2]. Later materials like steel, titanium, nickel were also being used by either using subtractive or additive manufacturing processes [3]. Towards the last two decades, microneedles fabricated out of polymers received an edge over other materials because they offered a lot of flexibility of manufacturing and customizable mechanical properties [4,5]. The materials used for microneedle fabrication range from metals, silicon, glass, ceramics and polymers like PLGA and SU-8 [3]. During drug administration through microneedles, there are chances that the microneedle tip might break after insertion in skin or during microneedle retraction. These broken tips should not cause bad immunogenic effects [6,7]. Hence, biocompatibility of the microneedle material remains the prime concern [8]. In spite of its wide use in earlier days, the biocompatibility of silicon is debatable [6,9]. The microneedle material should also be chemically resistant. All in all, the microneedles should be made strong enough to be able to address extreme conditions.

In polymers, SU-8, is often selected as a potential material for microneedle fabrication based on high crosslinked strength, biocompatibility, low-cost, light-induced polymerization and microelectronic industry process compatibility [10,9]. In Chapter 6, hollow SU-8 microneedles were fabricated and characterized for their mechanical strength and flowrate. This was done in order to test their efficacy for skin insertion and drug delivery. It was seen that SU-8 hollow microneedles were stronger by an order of magnitude to pierce human skin. This kind of strength is good enough for careful professional handling, but when the microneedles roll out as products in market and reach people where skilled medical force is scanty, then the microneedles

DOI: 10.1201/9781003202264-7

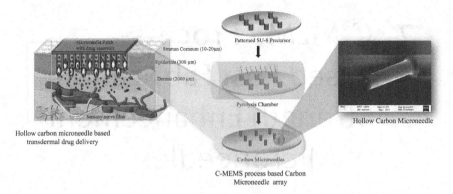

FIGURE 7.1 Graphical abstract of the chapter wherein the hollow microneedles puncture the skin to reach above the pain sensing nerves in transdermal region of skin painlessly and release drug by suitable actuation methods. It is followed by summary of the C-MEMS process followed and finally, the figure in outset shows the zoomed view of the single microneedle structure fabricated out of glassy carbon.

have to be lot stronger. They should be designed to handle stresses due to amateur handling, transportation and storage.

To solve this problem, this chapter introduces a new material for microneedle fabrication which is glassy carbon. Figure 7.1 shows the graphical abstract of the chapter.

Upon reading this chapter the reader will be able to

- Understand the structure and properties of glassy carbon.
- Gain insight about the carbon-microelectromechanical systems (C-MEMS) process of microfabrication.
- Know about the fabrication of hollow glassy carbon microneedles by (C-MEMS) process. This process acts as a post processing step for the previously discussed SU-8 microneedles which transforms the microneedle material to achieve desired microneedle properties.
- Understand the characterization techniques that establish superior mechanical properties of glassy carbon microneedles over SU-8 microneedles.

7.2   GLASSY CARBON – A PROMISING CANDIDATE FOR MICRONEEDLE FABRICATION

Carbon is one of the most interesting materials used for various applications including medical technology. A major advantage of choosing carbon over silicon or other materials is that it is available in various allotropes with unique physico-chemical properties which allows researchers to tailor its properties for unique applications [11]. The use of carbon in a medical field is not new. It has long been used for orthopaedic joints, carbon fibers and its composites are used for orthopaedic surgeries, surgical instruments, and for many other applications [12–15]. Further, they have been studied for the applications like artificial heart valve and ear drum repair material [16].

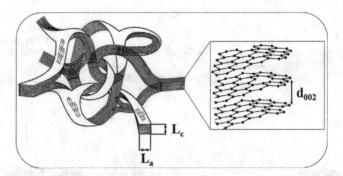

FIGURE 7.2 Glassy carbon structure.

Source: Courtesy de Souza Vieira et al. 2020 [22].

Carbon nanofibers reinforced composite have become new promising biomaterials and also aid in cancer treatment [17]. Also, three-dimensional scaffold-like structure generation from carbon nanofibers ensures bone tissue regeneration [18]. Carbon Nanotube (CNT) is another promising biomaterial with safe clinical use [19].

On the one hand we find synthetic carbon forms like fullerene, while on the other we find graphene which is one of the most researched allotrope of carbon. It is important to understand the structure of glassy carbon in order to appreciate its properties. Most common descriptions of glassy carbon describe it as interconnected grapheme ribbons with space between them or cage-like graphene structures similar to fullerene [20]. As per Pesin [21], the sp^2 hybridized carbon are arranged as graphitic planes stacked in disorganized fashion. This kind of arrangement is known as turbostratic arrangement. Such arrangement of glassy carbon is shown in Figure 7.2.

7.3 CARBON-MICROELECTROMECHANICAL SYSTEMS (C-MEMS)

Often a top-down approach is adopted to produce carbon devices. That is by patterning the structures using lithography and then carbonizing these structures. Such a process is known as C-MEMS process. The process steps are sown in Figure 7.3.

The material used prior to carbonization is known as a precursor. The C-MEMS process includes: spin coating of a carbon precursor photoresist, soft bake, UV exposure, post bake, and development [11,8]. After photolithography, carbon structures are obtained through a pyrolysis process. Pyrolysis involves thermochemical decomposition of polymers.

1. At temperatures below 550°C, first a carbon backbone structure is formed which acts as base for glassy carbon which forms in subsequent stages.
2. Pyrolytic carbon is obtained between temperatures 550°C–700°C. In this stage active radicals are detected to be present owing to the high number of dangling bonds.
3. At temperatures above 700°C, C-C bonds are formed. This stage is marked by development of short-range order and increase in crystallinity.

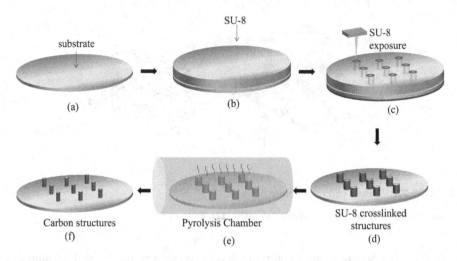

FIGURE 7.3 Typical steps of a C-MEMS process.

At temperatures more than 900°C, further graphitization occurs. At this stage most of the voids and rings are annealed. Even at such high temperatures some defects remain [20]. Resistant materials like phenol formaldehyde are well known to be used as carbon precursors. Several commercially available photoresists like SU-8 and AZ series photoresists can be used for patterning the precursors. The type of carbon structures obtained depends on the chemical composition of the photoresist. Hence it is very crucial to select a photoresist which will yield glassy carbon structure suitable for desired application [23].

Recent works in carbon-microelectromechanical systems (C-MEMS)-based glassy carbon microneedles have provided a milestone for biocompatible, high strength and patternable carbon microneedles [24].Conversion of SU-8 polymer precursor to glassy carbon microneedles by pyrolysis paves the way for high strength microneedles by carbon which is the building block of organic life and by the process which is low cost, reliable and simple [25–27]. Schueller et al. first demonstrated a practical carbon microfabrication technique using organic polymer precursors [28]. Since then C-MEMS-based structures have been used for a variety of applications, including batteries [29], fuel cells [30], dielectrophoresis [31], supercapacitors [32], scaffolds [33], carbon nanowire [34], transmission electron microscopy grids [35], Electroosmotic micropump [36] and gas sensors [37]. We shall now discuss the fabrication of hollow glassy carbon microneedles which is a novel approach for transdermal drug delivery and was earlier taken up by us [38].

7.3 DESIGN OF MICRONEEDLE

It is important to consider the appropriate dimensions of the microneedle for the reasons discussed in earlier chapters. In the case of glassy carbon microneedles, one needs to remember that upon pyrolysis, structural shrinkage leads to change in microneedle dimensions. Hence, microneedles of different geometries were

TABLE 7.1
Different Geometries of Fabricated Microneedles

Sample name	Outer diameter SU-8 microneedle	Inner diameter SU-8 microneedle	SU-8 microneedles wall thickness
A	100	40	30
B	100	50	25
C	100	60	20
D	100	70	15
E	100	80	10
F	100	90	5

considered for a parametric study on their dimensions. The overall shape remained a simple cylindrical hollow geometry. We considered 6 different designs for this work keeping the outer diameter constant to 100 μm for all of the designs and varying the the inner diameter from 40 μm to 90 μm with a 10 μm increment. A list of the dimension parameters is presented in Table 7.1.

7.4 FABRICATION OF HOLLOW GLASSY CARBON MICRONEEDLE

The C-MEMS process was adopted for the fabrication of carbon microneedles by patterning the SU-8 microneedles using direct laser writing technique [38] and convert them to glassy carbon microneedles upon pyrolysis. The flowchart of the required steps is shown in Figure 7.2. The process can mainly be categorized into following steps.

a) Fabrication of microfluidic ports
b) Fabrication of hollow SU-8 microneedles
c) Pyrolysis of SU-8 structures to yield glassy carbon microneedles

These steps are discussed in detail below.

7.4.1 FABRICATION OF MICROFLUIDIC PORTS

The microfluidic ports were etched in silicon wafer as first step similar to earlier work [38] as shown in Figure 7.5. For this, a 2" silicon (p-type, 100 plane) wafer was used as substrate and cleaned by the Piranha method (H_2SO_4 and H_2O_2 in the ratio 1:1) for 30 minutes and blown dry with nitrogen. Then after dehydration it was bake in oven for 30 minutes, about a 1 μm-thick SiO_2 layer was formed on the silicon by dry-wet-dry oxidation technique in a Tempress Oxidation furnace (Netherlands). The wafer was spincoated with negative photoresist (PR) HNR120 (Fujifilm, Japan) at 500 rpm for 5 seconds and at 3000 rpm for 20 seconds. After a prebake step for 30 minutes in the oven at 90⁰C, it was exposed in Karl Suss MA6 (Italy) mask aligner (UV 365 nm wavelength) for 6.5 seconds for pattern containing square pattern for potassium

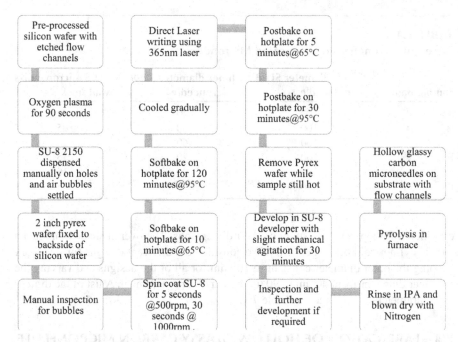

FIGURE 7.4 Process steps for fabrication of glassy carbon microneedles using SU-8 precursor by pyrolysis. The SU-8 microneedle structures shrink while retaining their overall geometry.

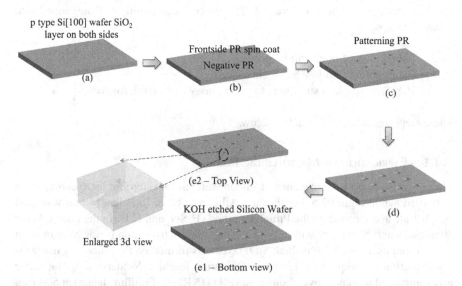

FIGURE 7.5 Process steps for fabrication of flow channels in silicon wafer using a wet etching technique.

hydroxide (KOH) etching of silicon. After the exposure, it was developed in WNRD developer (Fujifilm, Japan) and rinsed with n-butyl-acetate. It was postbaked at 120°C for 30 minutes giving a resist pattern on one side of the wafer. A similar procedure was carried out to pattern the second side of wafer with 400 μm squares aligned to the other side pattern. Silicon dioxide layer was removed from the unprotected region by etching with buffered hydrofluoric acid (BHF) for 4–5 minutes. Piranha solution (H_2SO_4 and H_2O_2 in the ratio 1:1) was used to remove the crosslinked photoresist layer in 30 minutes. Then KOH (40% by weight) solution was prepared at 70°C and flow channels were etched in silicon with the etch rate of 0.9 μm per minute. The sample was rinsed and dried. It was then slowly placed on 2" pyrex wafer to prevent SU-8 spilling from etched holes in subsequent steps and also to provide mechanical strength to the sample during handling.

7.4.2 FABRICATION OF SU-8 PRECURSOR MICROSTRUCTURES BY DIRECT LASER WRITING

Figure 7.6 shows the process steps for the hollow SU-8 precursor structure fabrication. SU-8 2150 (MicroChem, USA) was spun first at 500 rpm for 10 seconds to evenly spread the viscous photoresist and then at 1000 rpm for 30 seconds to get the desired thickness of 500 μm. The samples were then softbaked on hotplate at 65°C for 10 minutes and 95°C for 120 minutes. A highly coherent laser beam from a laser writer system (Helium-Cadmium laser at 325 nm, Microtech LW-405B, Italy, laser spot size 5 μm) was used for photo-polymerization of SU-8 at an intensity of 1277 mJ/ cm^2. The tightly focused Helium-Cadmium laser beam (365 nm) used for DLW has lesser diffraction related losses than UV photolithography technique and also avoids the problem of undesired SU-8 cross linking [39,10]. The alignment of the hollow

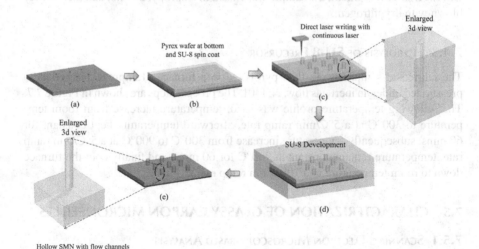

FIGURE 7.6 Process steps for hollow SU-8 microneedle structure fabrication. These structures are used as precursors for conversion to glassy carbon structures.

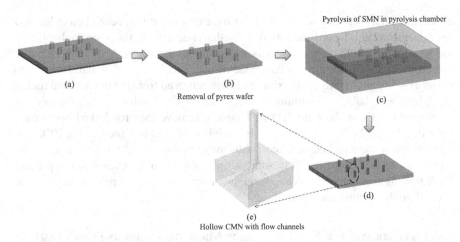

(a)

(b)

Removal of pyrex wafer

Pyrolysis of SMN in pyrolysis chamber

(c)

(d)

(e)

Hollow CMN with flow channels

FIGURE 7.7 Process steps for fabrication of glassy carbon microneedles from SU-8 precursor.

microneedles was done on the flow channels on substrate using alignment technique of the laser writer. In this technique, two pattern points (A&B) are searched on the wafer whose design coordinates are known. Once these points are marked, then the design file orients/ rotates itself to match the design points A&B to actual pattern points A&B on the wafer and the laser exposure happens in the SU-8 layer exactly above the patterns on the etched substrate. Post exposure baking was carried out for 5 minutes at 65°C and for 30 minutes at 95°C on a hotplate. After the postbake step, the pyrex wafer was removed while SU-8 is still hot and molten, and the sample was cooled down gradually. It was developed in the SU-8 developer solution (MicroChem, USA). After development, the sample was rinsed in isopropyl alcohol and then slowly blown with dry nitrogen.

7.4.3 Pyrolysis of SU-8 Precursor

The fabricated samples were then pyrolyzed in a furnace (Tempress, Netherlands) pressurized under an inert gas flow, N_2 [40]. The process steps are shown in Figure 7.7. The following temperature profile was used: temperature increase from room temperature to 300°C at a 5°C/min ramp rate, afterward temperature kept constant for 60 mins, subsequently temperature increase from 300°C to 900°C at a 5°C/min ramp rate, temperature remain constant at 900°C for 60 mins, and finally cool the furnace down to room temperature at a 10°C/min ramp rate.

7.5 CHARACTERIZATION OF GLASSY CARBON MICRONEEDLES

7.5.1 Scanning Electron Microscopy-based Analysis

Scanning electron microscopy (SEM) uses electrons in place of light for imaging. It allows a large amount of the sample to be in focus at one time and produces an image that is a good representation of the three-dimensional sample. It has high resolution and a lot of analytical instruments like electron probe microanalysis may also be added to a SEM system. Some of the advantages of SEM that make it one of the most useful instruments in various fields of research are mentioned below.

(a) high magnification,
(b) large depth of field,
(c) high resolution, and
(d) compositional and crystallographic information.

Figure 7.8 shows the SEM setup for the microneedle imaging. Figure 7.8a shows the setup for the surface topology and electron dispersive X-ray-based studies. Figure 7.8b shows that the sample stage that should be tilted by 25° to 45° so that cross sectional image of microneedles can be obtained. Figure 7.8c shows the SEM image setup where microneedles are cut through lumen so that alignment of microneedles on flow channels could be seen. This sample cut extends from microneedle to the flow channels in substrate.

SEM (EVO 18, ZEISS, Germany) was used to determine the morphology and cross-sectional profile of the microneedles. SEM micrographs were obtained for single SU-8 microneedle (precursor) and the corresponding pyrolyzed carbon microneedle structure and their dimensions were noted. The dimensions of SU-8 microneedle and corresponding dimensional change observed in carbon microneedle for different microneedle geometries is reported in Table 7.2.

The values of the different microneedle dimensions have been calculated from the SEM images. The corresponding SEM images of the samples showing the top view and cross-sectional view for samples is shown in Figure 7.9.

If we consider the microneedle of sample B shown in Figure 7.9, the outer diameter of 100 µm shrunk to 59.77 µm while the inner diameter 50 µm shrunk to 25.41 µm after pyrolysis. The microneedle wall thickness reduced from 25 µm to

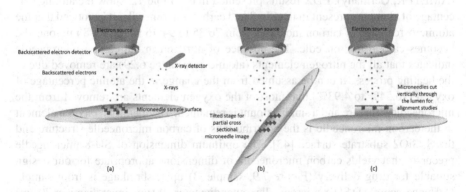

FIGURE 7.8 Scanning electron microscopy setup for (a) surface topology and electron dispersive X-ray-based studies, (b) cross sectional image of microneedles and (c) alignment of microneedles on flow channels in substrate.

TABLE 7.2
Shrinkage of Microneedle Dimensions upon Pyrolysis

Sample name	OD SMN	OD CMN	LD ratio	ID SMN	ID CMN	LD ratio	SW thickness	CW thickness	LD ratio
A	100	59.77	0.4	40	14.84	0.62	30	22.46	0.25
B	100	58.98	0.41	50	25.44	0.49	25	16.77	0.32
C	100	56.64	0.43	60	25.78	0.57	20	15.43	0.22
D	100	53.91	0.46	70	34.8	0.5	15	9.55	0.36
E	100	52.17	0.47	80	38.28	0.5	10	6.94	0.306
F	100	53	0.47	90	38.67	0.57	5	4.66	0.06

SMN – SU-8 microneedle, CMN – Carbon microneedle, OD – Outer Diameter, LD – Lateral decrease, ID – Inner Diameter, SW – SU-8 microneedle wall, CW – Carbon microneedle wall

16.77 µm. On average, there is 43% shrinkage in outer diameter while there is 54% shrinkage in inner diameter. The microneedle wall thickness is reduced on pyrolysis by an average 29%.

Along with SEM, there are various other signals which can provide us with a lot more information about the sample being analyzed. Electron dispersive X-ray (EDX) is one such technique. EDX is an X-ray-based technique used for analyzing elemental composition of a substance [41]. When an electron beam hits the inner shell of an atom, it knocks off an electron from the shell leaving behind a positively charged hole. As an electron moves from higher energy shell to occupy this empty level, the difference of energy in the levels is radiated in form of X-Ray. This energy is characteristic of specific element and energy level. The emitted X-ray are detected by a detector. This non-destructive method gives element as well as percentage composition in the substance [42].

Material composition was carried out using energy dispersive X-ray analysis (EDX) (AMETEK, Germany). EDX results presented in the Table 7.3 show the atomic percentage of elements present in the SU-8 and carbon microneedle. It is noticed that the atomic percentage of carbon increases from 76.35 to 94.46 when SU-8 microneedle becomes carbon microneedle. The absence of nitrogen in the carbon microneedle indicates that all the nitrogen elements (atomic percentage 6.29) are removed due to the heating process. It can be assumed from the change in the atomic percentage of oxygen (16.54% to 4.93%) that most of the oxygen elements are removed from the microneedle structure and a small amount is remained. The identified oxygen element in the carbon microneedle is the contributions of carbon microneedle structure and the Si/SiO2 substrate surface [43]. The optimum dimension of SU-8 microneedle precursor that yields carbon microneedle of dimensions appropriate for our design suitable for drug delivery (Figure 7.9, sample B) upon shrinkage is from sample D. Hence, sample D SU-8 microneedle (outer diameter 100 µm, inner diameter 70 µm) and corresponding carbon microneedle (outer diameter around 55 µm, inner diameter 35 µm) were used for all comparative characterizations discussed henceforth.

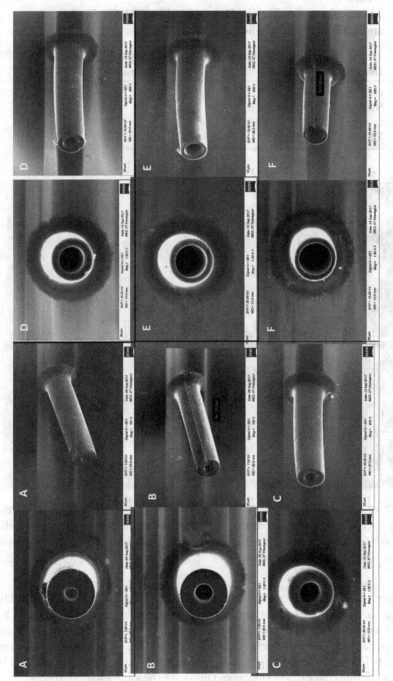

FIGURE 7.9 Scanning electron micrograph for glassy carbon microneedles corresponding to samples and corresponding tilted image to give an idea about the microneedle crosssection.

TABLE 7.3

Chemical Composition as Determined by EDX for SU-8 and Carbon Microneedle

Element	SU-8 microneedle Atomic %	Carbon microneedle Atomic %
Carbon	76.35	94.46
Nitrogen	6.29	-
Oxygen	16.54	4.93
Silicon	0.82	0.61

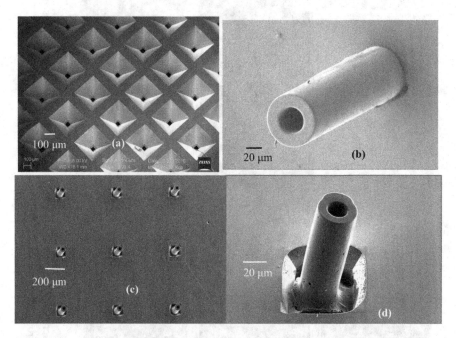

FIGURE 7.10 (a) Etched microfluidic conduit in silicon from where the drug shall flow from drug reservoir to carbon microneedle. (b) SU-8 microneedle fabricated on microfluidic conduit (backside of figure shown in a). (c) carbon microneedle array formed after pyrolysis. (d) zoomed view of carbon microneedle.

In order to fabricate the microneedle array as a device which can be readily connected to the drug reservoir and suitable actuation method, they should be fabricated on substrate with flow channels. Figure 7.10a shows the microfluidic conduits etched in silicon by wet chemical etching. Then the SU-8 microneedles were fabricated by direct laser writing which aligned them exactly on the etched conduits (40 μm on microneedle side) in silicon and these are shown in Figure 7.10b.

The pyrolysis process was carried out on the SU-8 microneedles and the resulting carbon microneedles are shown in Figure 7.10c. Figure 7.10d shows the zoomed view

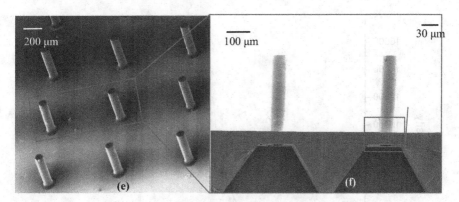

FIGURE 7.11 Optimized carbon microneedles aligned on etched microfluidic ports in silicon wafer, zoomed view of carbon microneedle aligned on flow channel in silicon.

of the carbon microneedle fabricated on the etched silicon substrate. In the process of conversion of SU-8 microneedle to carbon microneedle, the microneedles have almost detached themselves from the substrate. Hence the dimensions of the flow channel were reduced keeping in view the shrinkage ratio, so that the carbon microneedles cover the flow channel to avoid any fluid leakage or microneedle breakage. Therefore, on optimizing the flow channel dimensions, SU-8 microneedle (parameters of sample D) yielded carbon microneedles which covered the fluidic channels completely (Figure 7.11a). The zoomed view of the carbon microneedle and underlying flow channel is shown in Figure 7.11b–c.

7.5.2 RAMAN SPECTROSCOPY-BASED RESULTS

In the simplest terms, Raman microscopy uses light to excite molecular vibrations in a sample. These vibrations are then studied to analyze the sample chemically. It uses a monochromatic light source to irradiate matter. This causes inelastic scattering of light with a slight change in wavelength. This effect, where a small part of light changes its wavelength, is known as Raman scattering. The scattered light is analyzed with a spectrograph. A Raman spectrum is produced which consists of bands which are unique to certain functional groups and also for substances. This spectrum can provide information about composition, crystallinity and polymorphism. Raman microscope is a laser-based device which performs Raman microscopy [44].

Material composition and crystallinity were carried out using Energy dispersive X-ray analysis (EDX) (AMETEK, Germany) and Raman spectroscopy (Renishaw InVia Raman microscope, UK). Raman spectroscopy results of the microneedles are presented in Figure 7.12. Two peaks corresponding to the D- and G-band are centered around 1350 cm^{-1} and 1594 cm^{-1}, respectively. Earlier works have also reported the first-order Raman spectra which display two main peaks between 1200 cm^{-1} and 1700 cm^{-1}. These peaks are characteristic features of graphitic carbons [45]. The peak around 1600 cm^{-1} is called graphitic or G-band which arises due to optical phonon mode usually associated with in plane stretching of sp^2 bonded carbon atoms. The

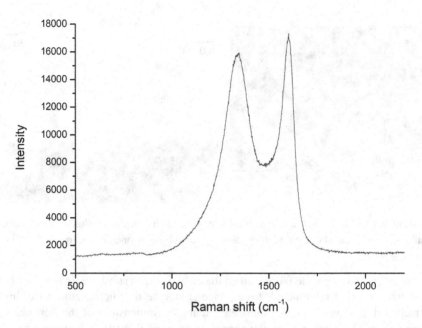

FIGURE 7.12   Raman spectra of glassy carbon microneedle.

peak around 1350 cm^{-1} is called disorder induced or D-band and is attributed to double resonant Raman scattering.

Crystallinity in the structure is defined by the intensity ratio (I_D/I_G) of these bands. This ratio is used to calculate the average in plane size of graphitic domains. The motive of this characterization was to check the nature of the obtained carbon. The results were matched with previous works and it was found that the microstructure of the carbon obtained here was glassy in nature.

7.5.3   NANOINDENTATION-BASED MECHANICAL CHARACTERIZATION

Hysitron Triboindentor TI 950 (Bruker, US) was used to characterize the mechanical properties of SU-8 and carbon microneedles. This measurement was performed using a Berkovich shaped indentor. The typical load displacement data for Berkovich indentor, as it was driven into and pulled out, into the SU-8 and carbon microstructures, were noted. The results are shown in Figure 7.13. Hardness and modulus of elasticity were calculated using the Oliver Pharr model. The hysteresis between the loading and unloading curve indicates the energy dissipated by indentation. Indentation performed on SU-8 microneedle (outer diameter 100 μm and inner diameter 40 μm) indicates the hard elasto-plastic behaviour with a hysteresis loop in between the loading and unloading curves. Quantitative analysis of the nanoindentation tests reveals a hardness of 0.33 GPa and Young's modulus of 5.52 GPa for carbon microneedle. Upon conversion of SU-8 microneedle to carbon microneedle on pyrolysis, the indentation curve changes to highly elastic material. Hardness increases to 2.62 GPa (8 times) and Young's modulus to 26.97 GPa (4.8 times).

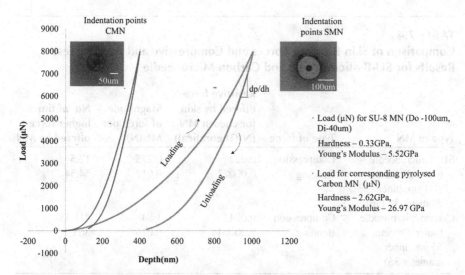

FIGURE 7.13 Load vs displacement data for SU-8 microneedle and corresponding pyrolyzed carbon microneedles.

7.5.4 Force Displacement-Based Mechanical Characterization

In order for the hollow microneedles to be used as A biological interface for trans-dermal drug delivery, their maximum bending and compression forces should be higher than the skin insertion force, so that they do not break during skin insertion.

As we discussed in previous chapters, we know that as a microneedle is inserted, it experiences resistive forces by human skin, out of which compression and bending forces are dominant. In order to successfully penetrate human skin, the applied force should be greater than this opposing force. The force required for puncturing skin is given by

$$F_{Skin} = P_{Puncturing} A \qquad (7.1)$$

If we consider a simple hollow cylindrical geometry of carbon microneedle with outer diameter 55μm and inner diameter 20μm as then the cross-sectional area is given by

$$A = \pi \left(R_o^2 - R_i^2 \right) \qquad (7.2)$$

Where Ro and Ri are outer and inner radius respectively. The bending forces acting on microneedle during skin insertion are of lower magnitude (around 1/100 of compression forces offered by skin). The resistive force offered by skin to carbon microneedle (sample D) is given in Table 7.4.

The microneedle array was loaded on Instron Microtester (Instron, USA) (Figure 7.14a). A metal plate was driven towards the microneedles (both SU-8 microneedle and carbon microneedle) until they broke. For bending tests, the two

TABLE 7.4

Comparison of Skin Resistive Forces and Compressive and Bending Test Results for SU-8 Microneedle and Carbon Microneedle

Type of MN	Type of force	Resistive force offered by skin for chosen MN (N)(Theoretical)	Magnitude of force per MN(N)	No. of times higher force offered by MN
SU-8 microneedle (outer diameter 100 µm, inner diameter 35 µm)	Compression	0.022	0.275	12.5
	Bending	0.00022	0.012	54.54
Carbon microneedle (outer diameter 55 µm, inner diameter 35)	Compression	0.0044	1.84	418.18
	Bending	0.000044	0.016	363

sides of wafer were held securely by clamps. For the compression test, the sample was loaded on a 90° rotated T shaped structure made of aluminium which was held securely between the clamps and the microneedles array substrate was stuck by glue on this fixture. The compressive and bending tests were performed for 4 samples each of SU-8 microneedle (100 µm outer diameter and 70 µm inner diameter) and carbon microneedle (55 µm outer diameter and 35 µm inner diameter).

The compression and bending test setup are shown in Figure 7.10b–c respectively.

The needles will not break if the applied load is below the maximum compressive or bending force. As the z stage applies increased load, eventually the needles break. The sharp drop in the curves (Figure 7.15a–b) marks the fracture point [46].

In compression tests, when the microneedles faced the metal plate, the plate pressed and broke the 100(10X10) microneedles together. In the bending tests, the microneedles were perpendicular to the metal plate and it bent and broke 10 microneedles at once. Hence the recorded load was divided by 100 for compression tests and by 10 for bending tests. These tests were repeated to obtain an average data plot. In Figure 7.15b, there is a slope variation at 0.1 mm which might be attributed to slight initial bending of the aluminium fixture. The actual compression of the microneedles started from 0.1 extension onwards. It was seen that for bending tests, the carbon microneedles showed higher bending strength than SU-8 microneedles (about 1.33 times higher). A similar trend was seen for compression tests where carbon microneedles showed higher compressive strength (about 6.7 times higher) than their precursor. Also, these carbon microneedles are much stronger to overcome the resistive forces offered by skin and shall pierce skin successfully (Table 7.4).

7.5.5 BIOLOGICAL SKIN INSERTION-BASED CHARACTERIZATION

Further, biological insertion-based characterization for the needles inserting in mice skin were carried out on a customized platform. The protocol was approved under the

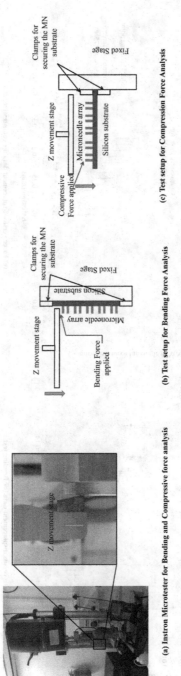

(a) Instron Microtester for Bending and Compressive force analysis

(b) Test setup for Bending Force Analysis

(c) Test setup for Compression Force Analysis

FIGURE 7.14 (a) Photograph of Instron Microtester for bending and compressive force analysis, (b) test setup for bending force analysis and (c) test setup for compression force analysis.

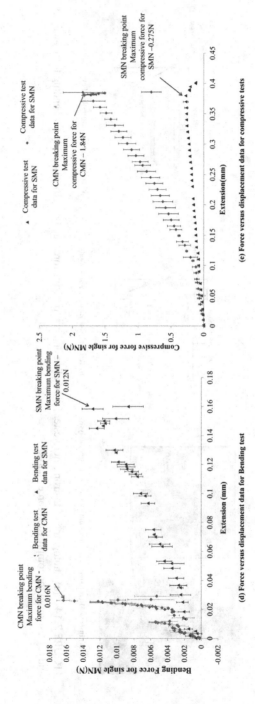

FIGURE 7.15 (a) Force vs displacement result of bending test and (b) force vs displacement result of compression test.

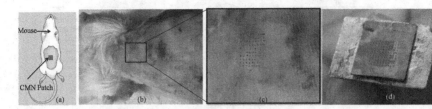

FIGURE 7.16 (a) Schematic of microneedle insertion test on mouse and (b) biological insertion test performed on 6–8 weeks Swiss Albino mouse. (c) Zoomed view of skin area where carbon microneedles have pierced and (d) intact array of 10X10 even after multiple insertions.

ethical clearance from the Committee for the Purpose of Control and Supervision of Experiments on Animals (CPCSEA), New Delhi, India. Mechanical interaction characterization of the hollow microneedles with mice skin was performed by driving the microneedle array perpendicularly against freshly excised mouse skin (Figure 7.16). The microneedle array was dipped in methylene blue and brought in contact with mouse skin, pressed against the skin and then slowly withdrawn. The procedure was repeated multiple times. On visual examination, the microneedle marks tainted in methylene blue could be seen faintly on the mouse. Some of the microneedle marks could not be seen due to uneven surface available for microneedle insertion. The spacing of microneedles is large enough for the absence of the "bed of nails" effect with 500 μm array spacing. The microneedle array was examined to see that all the microneedles remained intact after inserting the microneedle array at different places of mice skin even after 15 incisions (Figure 7.16c). The fabricated carbon microneedle with 500 μm array spacing successfully penetrated the mice skin surface and remained intact after retraction (Figure 7.16d). These tests on the microneedles substantiate the use of glassy carbon microneedles for transdermal drug delivery.

7.5.6 FLOWRATE CHARACTERIZATION

For flowrate measurements, the carbon microneedle array on silicon substrate was attached to a 5 mL syringe. A customized chamber containing DI water is used where pressurized nitrogen causes the flow of DI water to tube connected to syringe inlet. An Abus PT series pressure transducer monitored the pressure difference near the syringe inlet. In this setup, DI water can be passed through the carbon microneedle array at desired pressure difference. This water is collected in a calibrated beaker and volume dispensed over time is noted. This gives the flowrate of DI water through the carbon microneedles for a given pressure difference. The test setup for flow rate measurements of microneedles is shown in Figure 7.17a while the graph of flow rate versus pressure for individual microneedle is plotted in Figure 7.17b. Measurements were taken for 3 points corresponding to each pressure difference point and the average flowrate is plotted. It was noted that increased pressure accounts for increased flowrate while infusing more driving force. Overall, it showed that drug delivery could be controlled by controlling the inlet pressure. Hence known quantity

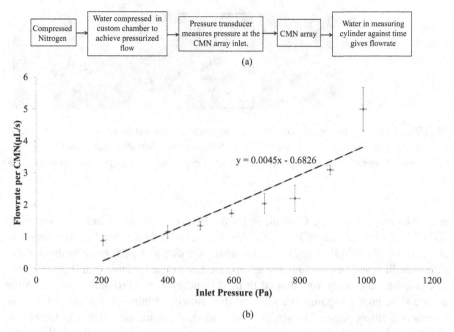

(a)

(b)

FIGURE 7.17 (a) Test setup for flow rate measurement and (b) plot of flow rate per microneedle at different inlet pressure.

of liquid drug formulation can be delivered to human body. For low pressure difference (<100 Pa), initial droplets issued out of carbon microneedles but on increasing the pressure, jet of water issued out of the carbon microneedles. The jet issuing out of the microneedles could be easily seen and almost all microneedles had jet of water coming out of them. This showed that 100% of microneedles were fabricated with clear lumen to allow water to flow through them.

7.6 SUMMARY

Carbon is an excellent choice of material for microneedle-based drug delivery applications. In this chapter, we learned of one fabrication approach for hollow carbon microneedles which is much harder than its precursor (SU-8) counterpart, thus, eliminating fear of any hazardous consequences arising out of microneedle tip breakage. A laser source was used to pattern the microneedle and pyrolysis process was considered for the conversion of SU-8 microneedle to glassy carbon microneedle. The whole conversion process was carried out at 1000°C temperature in N_2 atmosphere. A parametric study to freeze suitable dimensions for a microneedle to employ it for painless drug delivery. 500 μm was chosen as the microneedle length and 100 μm as outer diameter while the inner diameter was varied from 40 μm to 90 μm with 10 μm step. Structural shrinkage of the microneedle structure was observed and estimated using scanning electron microscope. It was observed that the hollow microneedle shape is retained in the structures. The EDX results provide the carbon atomic percentage in the microneedle structure is about 94. However, it also shows the presence of oxygen and silicon, those are due to substrate (Si/SiO$_2$) contribution. Raman spectroscopy results showed the microneedle structure is glassy in nature, which is biocompatible. The most important characterization is the quantitative analyses of hardness and Young's modulus using a nanoindentor. These analyses were carried out for both SU-8 and carbon microneedles and we estimated the increase of hardness and Young's modulus for the carbon microneedle was about 8 and 4.8 times respectively. A microneedle array was also inserted into freshly excised mice skin and withdrawn to examine the mechanical interaction of microneedle with skin. It was found that the microneedle penetrated the mice skin surface and remained intact after retraction multiple times. Once the microneedles are optimized and fabricated, we shall study the micropump part as the next component of the transdermal drug delivery system.

REFERENCES

1. R.F. Donnelly, T.R.R. Singh, D.I.J. Morrow, A.D. Woolfson, "*Microneedle Mediated Transdermal and Intradermal Drug Delivery*", Wiley, 2012.
2. J. Jing, E.H.T. Francis, M. Jianmin, I. Ciprian, "Microfabricated silicon microneedle array for transdermal drug delivery", *J. Phys.: Conf. Ser.* 34:1127–1131, 2006.
3. N. Roxhed, P. Griss, G. Stemme, "Membrane-sealed hollow microneedles and related administration schemes for transdermal drug delivery", *Biomed. Microdevices.* 10:271–279, 2008.
4. R.F. Donnelly, M.T.C. McCrudden, A.Z. Alkilari, E. Larranêta, E. McAllister, A.J. Courtenay, M. Kearney, T.R.R. Singh, H.O. McCarthy, V.L. Kett, E. Caffarel-Salvador, S.Al. Zaharani, A.D. Woolfson, "Hydrogel forming microneedles prepared from 'super swelling' polymers combined with hypophilised wafers for transdermal drug delivery", *PLOS ONE* 9, e111547, 2014.
5. F.Z. Rad et al., "High-fidelity replication of thermoplastic microneedles with open microfluidic channels", *Microsyst. Nanoeng.* 3(17034):1–11, 2017.
6. S. Henry, D.V. Mc Allister, M.G. Allen, M.R. Prausnitz, "Microfabricated microneedles: A novel approach to transdermal drug delivery", *J. Harm. Sci.* 87:922–925,1998.
7. K. Lee, C.Y Lee, H. Jung, "Dissolving microneedles for transdermal drug administration prepared by stepwise controlled drawing of maltose", *Biomaterials.* 32:3134–3140, 2011.
8. M. Madou, "*Fundamentals of Microfabrication*", Boca Raton, FL: CRC Press, 1997.
9. K. Kubo, N. Tsukasa, M. Uehara, Y. Izumi, M. Ogino, M. Kitano, T. Sueda, "Calcium and silicon from bioactive glass concerned with formation of nodules in periodontal ligament fibroblasts in vitro", *J. Oral Rehab.* 24:70–75,1997.
10. A. Del Campo, C. Greiner, "SU-8: A photoresist for high aspect ratio and 3D submicron lithography", *J. Micromech. Microeng.* 17:R81–R95, 2007.
11. C. Wang, G. Jia, L.H. Taherabadi, M.J. Madou, "A novel method for the fabrication of high-aspect ratio C-MEMS structures". *J. Microelectromech. Syst.* 14(2):348–358, 2005.
12. J.A Longo, J.B Koeneman, "Orthopedic applications of carbon fiber composites". In: D.L. Wise, D.J. Trantolo, K.U. Lewandrowski, J.D. Gresser, M.V. Cattaneo, M.J Yaszemski, (eds), *Biomaterials Engineering and Devices: Human Applications*, Totowa, NJ: Humana Press, 2000.

13. F.A. Sheikh, J. Macossay, T. Cantu, "Imaging, spectroscopy, mechanical, alignment and biocompatibility studies of electrospun medical grade polyurethane (Carbothane 3575A) nanofibers and composite nanofibers containing multiwalled carbon nanotubes", *J. Mech. Behav. Biomed. Mater.* 41:189–198, 2015.

14. S.A. Catledge, V. Thomas, Y.K. Vohra, "Nanostructured diamond coatings for orthopaedic applications". *Woodhead Publ. Ser. Biomater.* 2013, 105–150 2013.

15. T J Secker, R Hervé, Q Zhao, K B Borisenko, E W Abel, C W Keevil, "Doped diamond-like carbon coatings for surgical instruments reduce protein and prion-amyloid biofouling and improve subsequent cleaning". *Biofouling.* 28(6):563–9, 2012.

16. J.C. Bokros, "Carbon in medical devices", *Ceram. Int.* 9(1):3–7, 1983.

17. W.F. Xu, "Biocompatibility and medical application of carbon material", *Key Eng. Mater.* 452:477–480, 2011.

18. N. Saito, K. Aoki, Y. Usui, M. Shimizu, K. Hara, N. Narita, N. Ogihara, K. Nakamura, N. Ishigaki, H. Kato, H. Haniu, S. Taruta, Y.A. Kim, M. Endo, "Application of carbon fibers to biomaterials: A new era of nano-level control of carbon fibers after 30-years of development", *Chem. Soc. Rev.* 40:3824–3834, 2011.

19. N. Saito, H. Haniu, Y. Usui, K. Aoki, K. Hara, S. Takanashi, M. Shimizu, N. Narita, M. Okamoto, S. Kobayashi, H. Nomura, H. Kato, N. Nishimura, S. Taruta, M. Endo, "Safe clinical use of carbon nanotubes as innovative biomaterials", *Chem. Rev.* 114:6040–6079, 2014.

20. S. Sharma, C.N. Shyam Kumar, J.G. Korvink, C. Kübel, "Evolution of Glassy Carbon Microstructure: *In Situ* Transmission Electron Microscopy of the Pyrolysis Process", *Sci. Rep.* 8:16282, 2018 https://doi.org/10.1038/s41598-018-34644-9.

21. L.A. Pesin, Review: structure and properties of glass-like carbon, *J. Mater. Sci.* 37, 1–28, 2002.

22. L. de Souza Vieira, L.S. Montagna, J. Marini, F.R. Passador, Influence of particle size and glassy carbon content on the thermal, mechanical, and electrical properties of PHBV/glassy carbon composites. *J Appl Polym Sci.* 138:, e49740, 2021. https://doi.org/10.1002/app.49740

23. S. Sharma, "Glassy Carbon: A Promising Material for Micro–and Nanomanufacturing", *Materials* 11(10):1857, 2018 https://doi.org/10.3390/ma11101857.

24. B. Pramanick, S.O. Martinez-Chapa, M. Madou, "Fabrication of biocompatible hollow microneedles using the C-MEMS process for transdermal drug delivery", *ECS Trans.* 72(1):45–50, 2016.

25. B. Pramanick, S.O. Martinez-Chapa, M. Madou, H. Hwang, "Fabrication of 3D carbon microelectromechanical systems (C-MEMS)", *J. Vis. Exp.* 124:e55649, 2017.

26. H.W. Kroto, J.R. Heath, S.C.O. 'Brien, R.F. Curl, R.E. Smalley, "C60: buckminsterfullerene", *Nature* 318:162–163, 1985.

27. S. Ranganathan, R. McCreey, S.M. Majji, M. Madou, "Photoresist derived carbon for microelectromechanical systems and electrochemical applications", *J. Electrochem. Soc.* 147:277–282, 2000.

28. O.J.A. Schueller, S.T. Brittain, C. Marzolin, G.M. Whitesides, "Fabrication and characterization of glassy carbon MEMS", *Chem. Mater.* 9:1399–1406, 1997.

29. C. Liu, Y. Liu, M. Sokuler, D. Fell, S. Keller, A. Boisen, H. Butt, G.K. Auernhammer, E. Bonaccurso, "Diffusion of water into SU-8 microcantilevers", *Phys. Chem. Chem. Phys.* 12:10577–10583, 2010.

30. S. Holmberg, M.J. Madou, M. Rodriguez-Delgado, N. Ornelas-Soto, R.D. Milton, S.D. Minteer, R. Parra, "Bioelectrochemical study of thermostable pycnoporus sanguineus CS43 laccase bioelectrodes based on pyrolytic carbon nanofibers for bioelectrocatalytic O_2 reduction", *ACS Catal.* 5(12):7507–7518, 2015.

31. R. Martinez-Duarte, P. Renaud, M.J. Madou, "A novel approach to dielectrophoresis using carbon electrodes", *Electrophoresis* 32(17):2385–2392, 2011.
32. C. Kim, Y.O. Choi, W.J. Lee, K.S. Yang, "Supercapacitor performances of activated carbon fiber webs prepared by electrospinning of PMDA-ODA poly(amic acid) solutions", *Electrochim. Acta.* 50(2–3):883–887, 2004.
33. M. Blazewicz, "Carbon materials in the treatment of soft and hard tissue injuries", *Eur. Cell. Mater.* 2(2):21–29, 2001.
34. A. Salazar, B. Cardenas-Benitez, B. Pramanick, M.J. Madou, S.O. Martinez-Chapa, "Nanogap fabrication by joule heating of electromechanically spun suspended carbon nanofibers", *Carbon* 115:811–818, 2017.
35. B. Pramanick, A. Salazar, S.O. Martinez-Chapa, M.J. Madou, "Carbon TEM grids fabricated using C-MEMS as the platform for suspended carbon nanowire characterization", *Carbon* 113:252–259, 2017.
36. X. Wang, C. Cheng, S. Wang, S. Liu, "Electroosmotic pumps and their applications in microfluidic systems", *Microfluid Nanofluid.* 6(2):145, 2009.
37. Y. Lim, J.I. Heo, H. Shin, "Suspended carbon nanowire-based structures for sensor platforms", *ECS Trans.* 61(7):25–29, 2014.
38. T.K. Mishra, T.K. Maiti, T.K. Bhattacharyya, "Development of SU-8 hollow microneedles on silicon substrate with microfluidic interconnects for transdermal drug delivery", *IOP J. Micromech. Microeng.* 28:105017, 2018.
39. C. Decker, "Light induced crosslinking polymerization", *Polym Int.* 51:1141–1150, 2002.
40. R. Mardegana, S. Kamath, S. Sharma, P. Copeceb, P. Ugoa, M. Madou, "Optimization of Carbon Electrodes Derived from Epoxy-based Photoresist", *J. Electrochem. Soc.* 160(8):B132–B137, 2013.
41. M. Scimeca, S. Bischetti, H.K. Lamsira, R. Bonfiglio, E. Bonanno, "Energy Dispersive X-ray (EDX) microanalysis: A powerful tool in biomedical research and diagnosis", *Eur. J. Histochem.* 62(1):2841, 2018 https://doi.org/10.4081/ejh.2018.2841
42. www.thermofisher.com/blog/microscopy/edx-analysis-with-sem-how-does-it-work/ accessed on 06.09.21
43. B. Pramanick, M. Vázquez-Piñón, A. Torres-Castro, S.O. Martínez-Chapa, M. Madou, "Effect of pyrolysis process parameters on electrical, physical, chemical and electrochemical properties of SU-8-derived carbon structures fabricated using the C-MEMS process", *Mater. Today: Proc.* 5:9669–9682, 2018.
44. www.bruker.com/en/products-and-solutions/infrared-and-raman/raman-spectrometers/what-is-raman-spectroscopy.html accessed on 06.09.21
45. K. Jurkiewicz, M. Pawlyta, D. Zygadło et al., "Evolution of glassy carbon under heat treatment: correlation structure–mechanical properties", *J. Mater. Sci.* 53:3509–3523, 2018. https://doi.org/10.1007/s10853-017-1753-7
46. P.D. Shawn, J.L. Benjamin, H.A. Zachary, G.A. Mark, R.P. Mark, "Insertion of microneedles into skin: measurement and prediction of insertion force and needle fracture force", *J. Biomech.* 37:1155–1163, 2004.

8 Micropumps for Drug Delivery

8.1 INTRODUCTION

A pump is a device that moves fluid. In other words, it typically transfers electrical energy to mechanical energy. The earliest developed pumps were used to raise water and mostly employed in waterwheels. Then they were used in mining operation and developed as piston pumps. Energy is added to pumps using volumetric flows, electromagnetic force or the addition of kinetic energy [1]. Most of the pumps can be grouped as centrifugal pumps and positive displacement pumps. A centrifugal pump uses an electric motor which propels the impeller which in turn accelerates the fluid outwards. This type of pump introduces axial, radial or mixed flow in the fluid. On the other hand, positive displacement pumps move fixed amounts of fluids at regular intervals. They are filled up from internal cavities at suction and deliver a higher pressure of fluid at the outlet. Depending upon how the fluids are displaced, displacement pumps can be further classified as reciprocating pumps and rotary pumps. These pumps dominate the fluid movement in the macro world. After the world discovered the power of miniaturization in form or lab-on-chip devices, implantable or wearables, the demand for fluid manipulation at microscale increased. The small sized pumps called "micropumps", having fluid manipulation at microscale, were required for generating and maintaining a microfluidic flow. When it comes to manipulating small fluid volumes, macroscale physics no longer dominates and scaling laws come into the picture. Now the surface forces dominate the body forces. At microscale, laminar flows dominate instead of turbulent ones. Also, the capillary forces dominate over the viscous forces. Various types of micropumps have been developed over the years and for different applications. One of the rising needs from the biological and medical industry, which shall be discussed in this chapter is for drug delivery. Here the micropumps act as essential fluid-handling devices for precisely delivering liquid drug formulations in a specific direction. The chapter shall acquaint the reader with the different kinds of technologies that have been developed for micropumps and micropumps specifically for drug delivery.

8.2 MICROPUMPING TECHNIQUES

Micropumps are broadly categorized into mechanical displacement micropumps and electro- and magneto-kinetic micropumps [2].

1. Mechanical displacement micropumps – these micropumps employ oscillatory and rotational pressure on the fluid.

 DOI: 10.1201/9781003202264-8

2. Electro- and magneto-kinetic micropumps – these micropumps provide direct and continuous conversion of energy to pump fluid.

The above micropumps could be subdivided based upon their actuation principle. Let us look first at the different types of mechanical displacement micropumps.

8.2.1 Mechanical Displacement Micropumps

Mechanical micropumps vary widely in their design and parts but in general, we find the following universal components in all micropumps:

(a) Inlet with inlet valve – the inlet is the part of the micropump from which the fluid enters the pump chamber. The fluid flow is controlled by a valve. Often, a valveless design is also used to restrict fluid flow in one direction.
(b) Outlet with outlet valve – similar to inlet, the outlet valve allows directional fluid movement out of the micropump. One may choose a valve or a valveless design.
(c) Pump chamber – working fluid area.
(d) Actuator and diaphragm assembly.

8.2.1.1 Diaphragm Displacement Pumps

The diaphragm displacement pumps consist of inlet and outlet valves connected to a pumping chamber. The two modes of operation are expansion stroke and compression stroke. During expansion stroke, the diaphragm deflects decreasing chamber pressure. With the help of the inlet valve, the fluid fills the expanded pump chamber. The diaphragm then moves to decrease the chamber volume and increases the chamber pressure during the compression stroke. The fluid is discharged through the outlet valve. Diaphragm displacement pumps are further categorized on the basis of the actuation techniques they employ for moving the diaphragm to manipulate fluid flow. The broad categorization is given below. A comparison of the micropump actuators is presented in Table 8.1.

(a) Piezoelectric
(b) Thermal

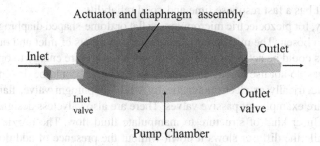

FIGURE 8.1 General structure of a diaphragm displacement pump.

TABLE 8.1
Comparison of Actuators Performance

Actuation mechanism	Force	Displacement	Response time	Reliability
Solenoid Plunger	Small	Large	Medium	Good
Piezoelectric	Very large	Medium	Fast	Good
Pneumatic	Large	Very small	Slow	Good
Shape Memory Alloy	Large	Large	Slow	Poor
Electrostatic	Small	Very small	Very fast	Very good
Thermopneumatic	Large	Medium	Medium	Good
Electromagnetic	Small	Large	Fast	Good
Bimetallic	Large	Small	Medium	Poor
Ionic EAP (IPMC)	Small	Large	Medium-fast	Good

    (c) Electrostatic
    (d) Electromagnetic / magnetic
    (e) Pneumatic
    (f) Composite / polymer materials
    (g) Peristaltic

8.2.1.1.1 Piezoelectric Micropumps

Piezoelectric actuators drive micropumps of this category. The piezoelectric actuator is deflected when subjected to an electric field, and vice versa. This is explained by the presence of non-symmetrical crystals that exist in a neutral balance. On application of pressure, the structure deforms, pushing the atoms around resulting in redistribution of charges which disturb the electrical neutrality of the crystal, leading to generation of an electric field. The reverse happens when an electric field is applied to a piezoelectric crystal; the atoms move to rebalance themselves and this leads to deformation. Quartz crystal, barium titanate and lead zirconate are examples of both naturally occurring and man-made piezoelectric crystals.

To employ the piezoelectric material in the micropump actuator, it may be bonded with, deposited on, or incorporated in the diaphragm membrane itself. Advantages of this type of actuation include relatively large forces and membrane displacement generated. It has a fast response time and good reliability.

Generally, for piezoelectric micropumps, a flat or dome-shaped diaphragm chamber geometry is chosen. The micropump chamber also consists of inlet and outlet valves. These valves could be active or passive. Active valves require energy to operate while passive valves do not use energy to operate. Active valves could be driven mechanically, piezoelectrically or electromagnetically while diaphragm valve, flap valve and ball valves are examples of passive valves. There are also valveless designs which use a nozzle diffuser kind of structure to manipulate fluid flow. The nozzle accelerates the fluid while the diffuser slows it down without the presence of additional flaps to achieve this. Multiple actuators can be connected if higher flow volume is desired.

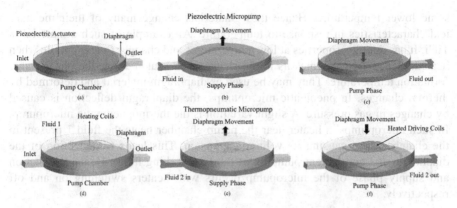

FIGURE 8.2   Different pumping phases for piezoelectric and thermopneumatic micropump, (a) piezoelectric micropump schematic, (b) supply phase and (c) pumping phase.

On application of external electrical actuation, the piezoelectric actuator exhibits many modes of vibration. One of the fundamental modes of vibration might be when the centre point of the piezoelectric disc undergoes maximum deflection. If used as a micropump actuator in this mode, the piezoelectric disc and membrane assembly will deflect upwards in one cycle of the applied voltage and deflect downwards in the next cycle. When the actuator assembly deflects upwards, low pressure is created in the chamber and the fluid from the reservoir enters the pump chamber. This is known as the **supply phase** of the micropump. When the actuator attached diaphragm deflects downwards in the pump chamber, it exerts pressure on the fluid present. Due to the valve design adopted in the pump, the flow of fluid from the inlet valve is restricted while from the outlet valve it is favoured. Hence, fluid comes out of the micropump through the outlet valve. This is known as the **pump phase**. The phases of piezoelectric micropump are shown in Figure 8.2a–c.

8.2.1.1.2   Thermal and Pneumatic Micropumps

Thermal micropumps work on the principle of diaphragm deflection due to volume expansion or induced stress in a material introduced by applied heat. The phenomenon of volume expansion is employed in a thermopneumatic micropump while thermal energy induced material stress is used in shape memory alloy-based micropumps. In a thermopneumatic micropump, a heating element heats a fluid (say fluid 1) which in turn expands the diaphragm (Figure 8.2d–f). This fluid 1 is different from the working fluid (say fluid 2). As the diaphragm expands, fluid 2 enters the pump chamber as in the suction cycle. Now the heater is switched off which cools the fluid 1 and it contracts. This deflates the diaphragm which causes it to contract and in turn forces fluid 2 out of the chamber.

The next kind of thermal micropumps are material expansion based. The most popular technique is the shape memory alloy-based actuator. The shape memory alloy-based actuators return to their predefined stage when they are subjected to a suitable heating procedure. When they cool, they are plastically deformed at some lower temperature. Hence these materials change many of their mechanical characteristics in response to temperature. An example of such a material is TiNi. It has ductile properties at low temperatures and changes to less ductile when heated. Hence TiNi assumes its initial shape when heated above the phase transformation temperature. They may be used as diaphragm material and deformed by thermal changes. In pneumatic micropumps, the diaphragm deflection is caused by changes in gas pressure. A slight variation is the thermopneumatic micropump. In such micropumps, a heater near the pump chamber heats the fluid 1 present in the chamber, thus causing its volume expansion. This leads to deflection of the diaphragm and consequent pumping of fluid 2 (working fluid). Hence the suction and supply phase of the micropump occurs with heaters switching on and off respectively.

8.2.1.1.3 Electromagnetic Micropumps

The electromagnetic micropumps consist of a permanent magnet attached to or embedded in the diaphragm. This magnet is surrounded by a coil and when current is passed through the coil electric and magnetic fields interact and produce Lorentz forces. This force is either attractive or repulsive and accordingly shall produce expansion or compression of the membrane. There are some interesting cases where even magnetic fluids are used to actuate the polymer diaphragm. These pumps have faster mechanical response but have large power consumption.

8.2.1.1.4 Electrostatic Micropumps

Electrostatic micropumps use the principle of electrostatic actuation where the electrostatic force between two oppositely charged electrodes moves the diaphragm membrane to which one of the electrodes is attached. These pumps consume low power. It is seen that the performance of such kinds of pump degrades over time due to a build-up of surface charges in the insulator region of the capacitor.

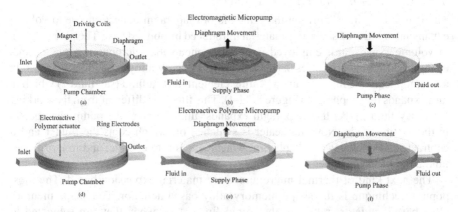

FIGURE 8.3 Different pumping phases for electromagnetic and electroactive polymer micropump, (a) piezoelectric micropump schematic, (b) supply phase and (c) pumping phase.

8.2.1.1.5 Electroactive Polymer Micropumps

Electroactive polymer composite materials were developed as diaphragm materials to achieve large deflection when actuated electromechanically. When a voltage is applied across the ionic polymer diaphragm electrodes, then two boundary layers, consisting of anions and cations, are created. The anion layers are fixed while the cations are movable. The cations dissociate from their cluster regions and rapidly migrate towards the cathode in response to the electric field. A bending moment is created when these hydrophilic cations drive the water molecules along with them, to the cathode. This results in accumulation of water molecules near the cathode and swelling of the membrane, thus causing deformation. After some time, water diffuses back to the membrane body inducing relaxation of the membrane [3,4,]. This phenomenon is utilized as an actuator to cause pump cycles. These micropumps achieve large deflection at low voltages, making them tough competitors of piezo-electric actuators.

8.2.1.1.6 Peristaltic Micropumps

In peristaltic micropumps, a pumping action is achieved by the sequential pumping of multiple actuators to achieve fluid flow in the desired direction. The actuation technique used may be piezoelectric, electromagnetic or any of the other actuation techniques previously discussed. Two types of peristaltic micropumps have been reported using piezoelectric discs and thermopneumatic actuators. The peristaltic micropump using piezoelectric discs was the first developed micropump.

8.2.2 ELECTRO- AND MAGNETO-KINETIC MICROPUMPS

The electro- and magneto-kinetic micropumps are also known as non-mechanical micropumps. These pumps use non-mechanical energy to drive fluids. The electro-kinetic micropumps use an electric field to push ions in the fluid, thereby pushing the entire fluid in the pump chamber. On the other hand, the magneto-kinetic micropumps use Lorentz force to drive the fluid. Lorentz force is the force exerted on a charged particle moving with velocity through an electric and magnetic field.

8.3 MICROPUMPS AS DRUG DELIVERY DEVICES

As we saw in the previous section, there are a variety of micropumps which could specifically release fluid at the desired rate and control fluid flow. This makes them ideal devices for drug delivery where they could deliver drugs at the desired rate and control infusion volumes. They could be made wearable or implantable. If the micropump is made wearable and the infusion volume is controlled, then it also eliminates the need of repeated needle injection. This also allows for long-term drug therapy like in the case of insulin management or cancer pain management.

Many times, micropumps are used to pump the drugs which otherwise find it difficult to penetrate the body [5–7]. Their advantages lie in the precise control of the drug flowrate (where they are human compliant either at fast or slow rates) and extremely small in size [8,9]. Micropump-based fluid delivery is either being used

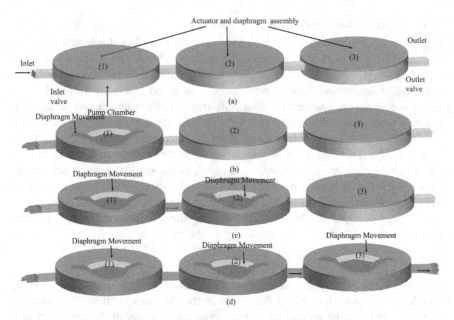

FIGURE 8.4 Schematic of peristaltic pump, (a) general structure consisting of three pump chambers and actuator assemblies. (b) sequential pumping of the peristaltic pump in which fluid enters pump 1, (c) outlet of pump 1 provides input for pump 2 which pushes fluid onwards to pump 3 and (d) pump 2 provides input for pump 3 which forces fluid out the micropump.

as a standalone device for drug delivery or is used in the lab-on-chip-based devices for diagnostics applications [10,11]. For example, for applications of drug delivery involving insulin delivery, flowrates of around 10–40 µL/min are required for bolus insulin delivery (i.e. insulin specifically taken at mealtimes to keep blood glucose level under control) [12,13]. Hence, the micropump targeting drug delivery should have low voltage operation, biocompatibility, portability, wide range of flowrate and light weight [14]. After the fabrication of the first MEMS-based micropump in the 1990s, a lot of micropump techniques have been investigated with the important parameters being actuation force, deformation, response time and reliability. [15–20].

8.4 MICROPUMP TARGET REQUIREMENTS

In order for the micropump to be used as a transdermal drug delivery device component, there are certain criteria one may look into before zeroing in down on the micropump specifications. The micropump should have dosage accuracy and precision. Systems with accurate (<1%) flow control are an invaluable solution. The system should check for the drug compatibility i.e. such structural materials should be chosen that minimize the possible loss of drug volume due to diffusion in the polymer matrix. It should be made from a biocompatible material so that it is biocompatible since it might come into contact with skin for long periods of time. It should not irritate the skin or cause allergic reactions to the patients. The power requirement is directly related to the device's lifetime. Hence micropump power consumption

TABLE 8.2
Micropump Requirements – Transdermal Insulin Delivery

Parameter	Application requirement	Choice for the case study example – transdermal insulin delivery
Micropump	Ease of fabrication	Valveless diffuser
Actuation	Large displacement	Piezoelectric
Flowrate	Precise controlled delivery of miniscule insulin doses particularly for infants	<0.5 μL/s
Controllable flowrate		Yes
Backflow		Minimized
Supply voltage	Application targeting human subjects	<30 V
Backpressure	Should be more than maximum human blood pressure to ensure drug delivery	>20 kPa
Operating temperature	Room temperature	

should be low. For successful transdermal drug delivery through the microneedles, the micropump must work against a nonzero backpressure of about 2 kPa [21], in addition to any subcutaneous pressure [22–24].

These requirements help us in defining the initial specifications of the device while designing a product for drug delivery. We may consider an example of a case of drug delivery device for transdermal insulin delivery which has micropump as its driving component. Then based upon our above discussion, we may work to arrive at the following initial specification of the device. These specifications are indicative only and vary widely upon application and requirement.

These requirements are summarized in Table 8.2.

8.5 DRUG RESERVOIR TARGET REQUIREMENTS

The specifications for one other component that needs to be decided along with micropump specifications are the reservoir requirements. We need to understand the different aspects of the drug reservoir to which the micropump will be attached. If the drug delivery application requires dosage which can be housed in the micropump chamber itself, then there is no need for a separate drug reservoir. But to deliver large volumes of drugs, an external reservoir needs to be designed along with the micropump.

The drug reservoir should ideally have the following features:

1. The reservoir capacity should depend on the required dose and drug potency.
2. The structural material for the reservoir should not pose an obstacle to the reservoir's mechanical compliance and limit its coupling methods to the

micropumping module. Hence PDMS is a well-suited elastomer for this process as it allows batch fabrication and forms leakage proof systems.

3. The fluid retention rate of the microreservoir should be high i.e. it should aid long-term storage capability.

8.6   VALVELESS DIFFUSER MICROPUMP – A CASE STUDY

We have seen that valveless diffuser pumps allow ease of fabrication and help in flow regulation. As a case study, we shall investigate this valveless micropump in detail and gain an in-depth understanding of the micropump theoretical analysis. As we know, the valveless diffuser pump uses diffuser elements as flow directing elements. The first valveless diffuser pump was presented in 1993. Wear and fatigue in the valves are eliminated since the diffuser elements have no moving parts and the risk of the valves clogging is also reduced. The diffuser pump is a positive displacement pump in the sense that it has a moving boundary that forces the fluid along by volume changes. As with other positive displacement pumps, it delivers a periodic flow. The pump principle has been shown to work for different liquids and for air.

8.6.1   DIFFUSER ELEMENT

The diffuser, a flow channel with gradually expanding cross section, is the key element in the valveless diffuser pump. Used in the opposite direction with a converging cross section it is called a nozzle. Diffusers usually have circular or rectangular cross sections (Figure 8.5). They are called conical and flat-walled diffusers, respectively. Both diffusers and nozzles are common devices in macroscopic internal flow systems.

The function of the diffuser is to transform kinetic energy, i.e. velocity, to potential energy, i.e. pressure. The type of flow in a diffuser can be exemplified by a "stability map", such as that shown in Figure 8.6. The map shows that depending on the diffuser geometry, the diffuser operates in four different regions. In the no-stall region the flow

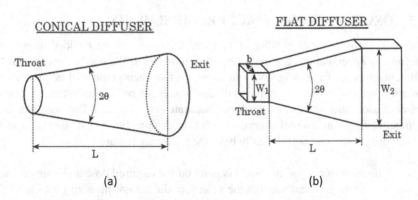

FIGURE 8.5   Different kinds of valveless diffusers (a) conical diffuser and (b) flat diffuser.

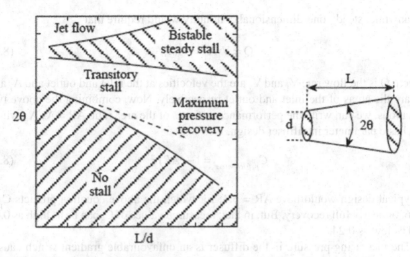

FIGURE 8.6 A stability map of a diffuser.

is steady viscous without separation at the diffuser walls and has moderately good performance. In the transitory steady-stall region the flow is unsteady. The minimum pressure loss occurs in this region. In the bistable steady region, a steady bistable stall can flip-flop from one part of the diffuser wall to the other and performance is poor. In the jet flow region, the flow separates almost completely from the diffuser walls and passes through the diffuser at nearly constant cross-sectional area making performance extremely poor.

Neglecting losses and gravity effects, the incompressible Bernoulli equation predicts that

$$p + \frac{1}{2}\rho V^2 = p_0 = const \qquad (8.1)$$

Where p_0 is the stagnation pressure which the fluid would achieve if the fluid were slowed to rest (V=0) without losses.

The basic output of a diffuser is the pressure recovery coefficient C_p defined as

$$C_p = \frac{p_e - p_t}{p_{0t} - p_t} \qquad (8.2)$$

Here the subscripts e and t mean the exit and the throat (or inlet), respectively. Higher C_p means better performance.

Now,

$$C_{p,frictionless} = 1 - (\frac{V_2}{V_1})^2 \qquad (8.3)$$

Meanwhile, steady one-dimensional continuity would require that

$$Q = V_1 A_1 = V_2 A_2 \tag{8.4}$$

Where Q is the flow rate V_1 and V_2 are the velocities at the inlet and outlet and A_1 and A_2 are the areas of the inlet and outlet respectively. Now, combining the above two equations, we can write the performance in terms of the area ratio $AR = A_2/A_1$, which is a basic parameter in diffuser design.

$$C_{p,frictionless} = 1 - (AR)^{-2} \tag{8.5}$$

A typical design would have AR = 5:1; for which the above equation predicts C_p = 0.96, or nearly full recovery. But, in fact, measured values of C_p can be as high as 0.86 and as low as 0.24.

The increasing pressure in the diffuser is an unfavourable gradient which causes the viscous boundary layers to break away from the walls and greatly reduces the flow performance. In fact, the shearing stresses at the fluid-solid interface cause the boundary layers to develop and in some places the velocity gradient perpendicular to the walls becomes zero and there is a local flow reversal. When flow visualization was possible this erratic flow patterns of a diffuser were revealed.

If one considers typical performance maps for diffusers then the higher the C_p value, the better is the diffuser performance. In general, the two main types of diffusers, conical and flat-walled, have approximately the same diffuser capacity. However, the best performance for conical diffusers is achieved at a length that is 10 to 80 per cent longer than for the best flat-walled design. The choice of diffuser type depends mainly on the fabrication process, but flat-walled diffusers are preferred since they give a more compact design.

The loss coefficient is related to the C_p value by the relation

$$K = 1 - \left(\frac{A_{throat}}{A_{exit}} \right)^2 - C_p \tag{8.6}$$

Where A_{throat} and A_{exit} are the throat and exit cross-sectional areas, respectively. For small angles the losses in the diffuser are small and the minimum losses occur for a cone angle 2θ equal to about 5°. For cone angles larger than 40° to 60° the loss is higher than for a sudden expansion. For these large angles, the gradual expansion does not raise the static pressure further and there is no diffuser effect. This unexpected effect is due to gross flow separation in a wide-angle diffuser. The effect is highly dependent on the inlet boundary conditions.

In order to achieve the best pump performance, the diffuser element has to be designed for the highest possible flow directing capability. To estimate the possible flow directing capability of a diffuser element, the available information for macroscopic internal flow systems with a circular cross section can be used. The pressure drop in an internal flow system is usually given as the loss coefficient, K, which is related to the pressure drop, Δp by

$$\Delta p = K \cdot \frac{1}{2} \rho \underline{u}^2_{upstream} \qquad (8.7)$$

where ρ is the fluid density and $\underline{u}_{upstream}$ is the mean velocity upstream. For the complete diffuser element, it is more practical to relate the pressure drop to the velocity in the narrowest cross section, the throat, as

$$\Delta p = \xi \cdot \frac{1}{2} \rho \underline{u}^2_{throat} \qquad (8.8)$$

where ξ is the pressure loss coefficient and $\underline{u}_{throat}$ is the mean velocity in the throat. The relation between ξ and K is then the simple area relation

$$\xi = K \cdot \left(\frac{A_{throat}}{A_{upstream}} \right)^2 \qquad (8.9)$$

With this definition the diffuser element efficiency ratio η can be defined as

$$\eta = \frac{\xi_{negative}}{\xi_{positive}} \qquad (8.10)$$

To optimize the efficiency of the diffuser element the ratio should be maximized. To achieve this, the entrance region of the diffuser should be rounded [25] and the outlet should be sharp.

8.6.2 Pressure Loss Coefficient

The pressure loss coefficient for flows through a gradually contracting nozzle, a gradually expanding diffuser, or a sudden expansion or contraction in an internal flow system is defined as the ratio of pressure drop across the device to the velocity head upstream of the device and is given by

$$K = \frac{\Delta p}{\dfrac{\rho v^2}{2}} \qquad (8.11)$$

For flow through a gradually expanding diffuser or a gradually contracting nozzle, the pressure loss coefficient can be calculated as follows. For flow in the diffuser direction (from cross section a to b in Figure 8.7), the incompressible steady-flow energy equation reduces to

$$P_a + \frac{1}{2}\rho v_a^2 = P_b + \frac{1}{2}\rho v_b^2 + \Delta p_d \qquad (8.12)$$

Hence the pressure loss coefficient can be written as

$$K_d = \frac{\Delta p_d}{\frac{\rho v_a^2}{2}} = \frac{P_a - P_b}{\frac{\rho v_a^2}{2}} + \left(1 - \frac{v_b^2}{v_a^2}\right) \qquad (8.13)$$

Introducing the pressure recovery coefficient

$$C_p = \frac{P_b - P_a}{\frac{\rho v_a^2}{2}} \qquad (8.14)$$

and using the continuity equation $A_a v_a = A_b v_b$, K_d for spatial diffusers (e.g. conical and pyramidal) can be written as

$$K_d = 1 - \frac{d_a^4}{d_b^4} - C_p \qquad (8.15)$$

since $A \propto d^2$. While for planar diffusers, $A \propto d$ and hence K_d is given by

$$K_d = 1 - \frac{d_a^2}{d_b^2} - C_p \qquad (8.16)$$

Hence, for a given diffuser geometry, the pressure loss coefficient can be calculated from the pressure drop and the mean velocity at the neck. Similarly, for flow in the nozzle direction (from cross section b to a in Figure 8.7), the pressure loss coefficient is given by

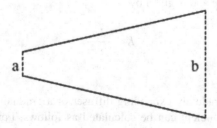

FIGURE 8.7   Schematic of a nozzle diffuser element.

$$K_n = \frac{\Delta p_n}{\frac{\rho v_b^2}{2}} \tag{8.17}$$

The diffuser efficiency of a nozzle diffuser element is defined as the ratio of the total pressure loss coefficient for flow in the nozzle direction to that for the flow in the diffuser direction

$$\eta = \frac{K_{n,t}}{K_{d,t}} \tag{8.18}$$

Hence, $\eta > 1$ will cause a pumping action in the diffuser direction in a valveless micropump, while $\eta < 1$ will lead to pumping action in the nozzle direction. The case where $\eta = 1$ corresponds to equal pressure drops in both the nozzle and the diffuser directions, leading to no flow rectification.

In the above equation, the total pressure loss coefficients for both the diffuser and nozzle directions can be divided into three parts:

i.   losses due to sudden contraction at the entrance,
ii. losses due to gradual contraction or expansion through the length of the nozzle diffuser, and
iii. losses due to sudden expansion at the exit.

The total pressure drop in the diffuser and nozzle direction can thus be written as

$$\Delta p_d = \Delta p_{d,1} + \Delta p_{d,2} + \Delta p_{d,3} \tag{8.19}$$

$$\Delta p_n = \Delta p_{n,1} + \Delta p_{n,2} + \Delta p_{n,3} \tag{8.20}$$

Where suffix 1, 2 and 3 denote regions 1, 2 and 3 respectively.

Using equation (8.7) and the continuity equation, equation (8.19) and (8.20) can be written as

$$\Delta p_d = \left[K_{d,1} + K_{d,2} + K_{d,3} \left(\frac{A_1}{A_3} \right)^2 \right] \frac{1}{2} \rho u_1^2 \tag{8.21}$$

$$\Delta p_n = \left[K_{n,1} + (K_{n,2} + K_{n,3}) \left(\frac{A_1}{A_3} \right)^2 \right] \frac{1}{2} \rho u_1^2 \tag{8.22}$$

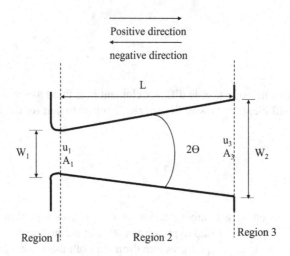

FIGURE 8.8 Schematic of nozzle/diffuser with three flow regions.

For the complete diffuser element, we now use ξ as the pressure loss coefficient instead of the previously used K and

$$\Delta p = \xi \cdot \frac{1}{2} \rho \underline{u}^2$$ (8.23)

Where $\underline{u}$ denotes the mean velocity in the diffuser neck. The total pressure loss coefficient in the diffuser direction can thus be written as

$$\xi_d = K_{d,1} + K_{d,2} + K_{d,3} \left(\frac{A_1}{A_3} \right)^2$$ (8.24)

And in the nozzle direction as

$$\xi_n = K_{n,1} + \left(K_{n,2} + K_{n,3} \right) \left(\frac{A_1}{A_3} \right)^2$$ (8.25)

So, the diffuser element efficiency ratio

$$\eta = \frac{\xi_n}{\xi_d} = \frac{K_{n,1} + \left(K_{n,2} + K_{n,3} \right) \left(\frac{A_1}{A_3} \right)^2}{K_{d,1} + K_{d,2} + K_{d,3} \left(\frac{A_1}{A_3} s \right)^2}$$ (8.26)

For example, let us consider the following parameters for nozzle diffuser element dimensions:

Diffuser length (L): 675 μm
Throat width: 45 (W1) μm & Area: (45×45) μm^2
Outlet width: 113 (W2) μm & Area: (113×45) μm^2
Depth (b) = 45 um
So, the area ratio (AR) = 2.51, L/W1 = 15

So, from the perf ormance map, divergence Angle (2θ): 5.72 degrees

$$and \quad C_p \cong 0.65$$

So, we have, $\xi_{positive} = \xi_{in} + \xi_{diverging} + \xi_{out}$
$= 0.4 + 0.2 + 0.15$
$= 0.75$

And, $\xi_{negative} = \xi_{in} + \xi_{converging} + \xi_{out}$
$= 0.06 + 0.0045 + 1$
$= 1.0645$

The diffuser element efficiency ratio

$$\eta = \frac{\xi_{negative}}{\xi_{positive}} = \frac{1.0645}{0.75} \cong 1.42$$

8.6.3 FLOW RECTIFICATION EFFICIENCY

The flow rectification efficiency of a valveless micropump is the measure of the ability of the pump to direct the flow in one preferential direction. It can be expressed as

$$\epsilon = \frac{Q_+ - Q_-}{Q_+ + Q_-} \tag{8.27}$$

in which Q is flow rate and subscripts + and − refer to flow in the forward and the backward directions, respectively. A higher ϵ corresponds to better flow rectification. In particular, when there is no flow rectification, equal amounts of fluid move in both directions and $\epsilon = 0$, while for perfect rectification, flow only in one direction and $\epsilon = 1$. The flow rectification efficiency of a valveless micropump is related to the diffuser efficiency of the nozzle diffuser elements. As the diffuser efficiency departs from a value of 1, i.e. as the difference between $K_{n,t}$ and $K_{d,t}$ increases, ϵ for the micropump also increases.

8.6.4 PUMPED VOLUME

The pressure loss coefficient for flows through a gradually contracting nozzle or gradually expanding diffuser or a sudden expansion or contraction in an internal flow system is defined as the ratio of pressure drop across the device to the velocity head upstream of the device. So, we have

$$\Delta P_d = \frac{\rho v_d^2}{2}\xi_d \quad \text{for diffuser,} \tag{8.28}$$

and

$$\Delta P_n = \frac{\rho v_d^2}{2}\xi_n \quad \text{for nozzle} \tag{8.29}$$

Where, ρ is the fluid flow density and v_d and v_n are the fluid flow velocities in the narrowest part (throat) of the diffuser and nozzle, v_d and v_n are assumed to be constant across the cross section; ξ_d and ξ_n are the pressure loss coefficient of the diffuser and nozzle respectively. The volume flows in the diffuser and nozzle are

$$\Phi_d = A_d v_d \text{ and } \Phi_n = A_n v_n \tag{8.30}$$

where, A_d and A_n are the cross-sectional areas of the narrowest parts of the diffuser and nozzle respectively.

So,

$$\Phi_d = A_d \left(\frac{2}{\rho}\right)^{\frac{1}{2}}\left(\frac{\Delta P_d}{\xi_d}\right)^{\frac{1}{2}}$$

$$\text{and } \Phi_n = A_n \left(\frac{2}{\rho}\right)^{\frac{1}{2}}\left(\frac{\Delta P_n}{\xi_n}\right)^{\frac{1}{2}} \tag{8.31}$$

Using the same element as diffuser and nozzle gives the same cross-sectional throat area:

$$A_d = A_n = A \tag{8.32}$$

The inlet and outlet elements are assumed to be identical. If the inlet and outlet pressure P_1 and P_0 can both be neglected compared to the chamber pressure P_c, the volume flows in the diffuser and nozzle direction can be written as

$$\Phi_d = \frac{C}{\xi_d^{\frac{1}{2}}} \text{ and } \Phi_n = \frac{C}{\xi_n^{\frac{1}{2}}} \text{ Where, } C = A\left(\frac{2P_c}{\rho}\right)^{\frac{1}{2}} \tag{8.33}$$

The chamber volume variation is

$$V_c = V_x \, sinsin(\omega t) \text{ with } \omega = 2\pi f \tag{8.34}$$

Where, V_x is the volume variation amplitude and f is the pump frequency. This gives a net chamber volume flow of

$$\Phi_1 - \Phi_0 = \frac{dV_c}{dt} = V_x \omega coscos(\omega t) \tag{8.35}$$

where Φ_1 is the volume flow into the chamber through the inlet and Φ_0 is the volume flow out of the chamber through the outlet.

During the supply mode the chamber volume increases, $dV_c/dt > 0$, which gives a net flow into the chamber with the inlet element acting as a diffuser and the outlet element acting as a nozzle. This gives inlet and outlet flows of

$$\Phi_1 = \Phi_d = \frac{C}{\xi_d^{\frac{1}{2}}} \text{ and } \Phi_0 = -\Phi_n = -\frac{C}{\xi_n^{\frac{1}{2}}} \tag{8.36}$$

This yields a net chamber flow of

$$\Phi_1 - \Phi_0 = C\left(\frac{1}{\xi_d^{\frac{1}{2}}} + \frac{1}{\xi_n^{\frac{1}{2}}} \right) = V_x \, \omega coscos(\omega t) \tag{8.37}$$

which gives

$$C = \frac{V_x \omega coscos(\omega t)}{\left(\dfrac{1}{\xi_d^{\frac{1}{2}}} + \dfrac{1}{\xi_n^{\frac{1}{2}}} \right)} \tag{8.38}$$

The supply mode outlet flow is

$$\Phi_s = -\Phi_n - \frac{C}{\xi_n^{\frac{1}{2}}} \tag{8.39}$$

which with the expression for C yields

$$\Phi_s = \frac{-V_x \omega coscos(\omega t)}{\left[1 + \left(\dfrac{\xi_n}{\xi_d} \right)^{\frac{1}{2}} \right]} \tag{8.40}$$

During the pump mode the chamber volume decreases, $dV_c/dt < 0$, which gives a net flow out of the chamber with the inlet element acting as a nozzle and the outlet element acting as a diffuser. This gives inlet and outlet flows of

$$\Phi_1 = -\Phi_n = -\frac{C}{\xi_n^{\frac{1}{2}}} \tag{8.41}$$

and

$$\Phi_0 = \Phi_d = \frac{C}{\xi_d^{\frac{1}{2}}} \tag{8.42}$$

So the pump mode outlet flow is

$$\Phi_p = \frac{-V_x \omega coscos(\omega t)}{\left[1 + \left(\dfrac{\xi_d}{\xi_n}\right)^{\frac{1}{2}}\right]} \tag{8.43}$$

Now we could calculate the pump volume as follows. If the pressure loss coefficients are assumed to be constant throughout the pump cycle, the total pumped volume during one complete pump stroke is

$$V_0 = \int_{-T/4}^{T/4} \Phi s + \int_{T/4}^{3T/4} \Phi p \tag{8.44}$$

$$= \frac{-V_x}{\left[1 + \left(\dfrac{\xi_n}{\xi_d}\right)^{\frac{1}{2}}\right]} \int_{-T/4}^{T/4} \omega coscos(\omega t)\,dt + \frac{-V_x}{\left[1 + \left(\dfrac{\xi_d}{\xi_n}\right)^{\frac{1}{2}}\right]} \int_{T/4}^{3T/4} \omega coscos(\omega t)\,dt \tag{8.45}$$

$$\Rightarrow V_0 = 2V_x \left[\frac{\eta_{nd}^{\frac{1}{2}} - 1}{\eta_{nd}^{\frac{1}{2}} + 1} \frac{\eta^{\frac{1}{2}} - 1}{\eta^{\frac{1}{2}} + 1}\right] \quad \text{Where,} \, \eta = \frac{\xi_n}{\xi_d} \tag{8.46}$$

8.7   CONCLUSION

This chapter acquainted readers with different kinds of micropump technologies, especially the ones used for drug delivery. It covered an in-depth understanding of the different part of the micropump and there theoretical modelling by taking the case study of valveless diffuser nozzle-based micropump. With this input, the readers can then proceed to equip themselves with a simulation-based understanding of micropump technologies.

REFERENCES

1. www.britannica.com/technology/pump accessed on 28.06.2022.
2. B.D. Iverson, S.V. Garimella, "Recent advances in microscale Pumping technologies: A Review and evaluation" CTRC Research Publications. Paper 89. http://dx.doi.org/10.1007/s10404-008-0266-8, 2008.
3. P. de Gennes, K. Okumura, M. Shahinpoor, K.J. Kim, "Mechanoelectric effects in ionic gels", *Europhys. Letters.* 50:513–518, 2000.
4. Y. Cha, M. Porfiri, "Mechanics and electrochemistry of ionic polymer metal composites", *J. Mech. Phys. Solids.* 71:156–178, 2014.
5. H.A.E. Benson, A.C. Watkinson, ed. *"Topical and Transdermal Drug Delivery: Principles and Practice"*, 6–7, Wiley, 2012. ISBN 978-0-470-45029-1
6. E.A. Tetteh, M.A. Boatemaa, E.O. Martinson, "A review of various actuation methods in micropumps for drug delivery applications," *2014 11th International Conference on Electronics, Computer and Computation (ICECCO)*, Abuja, 1–4, 2014.
7. C. Joshitha, B.S. Sreeja, S. Radha, "A review on micropumps for drug delivery system," *2017 International Conference on Wireless Communications, Signal Processing and Networking (WiSPNET)*, Chennai, 186–190, 2014.
8. P.K. Das, A.B.M.T. Hasan, "Mechanical micropumps and their applications: A review", *AIP Conf. Proc.* 1851(1), 020110, 2017.
9. Y.N. Wang, L.M. Fu, "Micropumps and biomedical applications–A review", *Microelect. Eng.* 195:121–138, 2018.
10. H. Genslar, R. Sheybani, P.Y. Li, R.Lo, E. Ming, "An implantable MEMS micropump system for drug delivery in small animals", *Biomed. Microdevices.* 14(3), 483–496, 2012.
11. K. Siebal, L. Scholer, H. Schafer, N. Bohm, "A programmable planar electroosmotic micropump for lab-on–chip applications", *J. Micromech. Microeng.* 18(2):025008, 2008.
12. R.R.V. Chemitiganti, C.W. Spellman, "Management of progressive type 2 diabetes: Role of insulin therapy.", *Osteopath. Med. Prim. Care.* 3(1):5, 2009.
13. www.sps.nhs.uk/wp-content/uploads/2018/.../Insulin-pump-table-May-2018.pdf accessed on 19.09.18.
14. AO.N. Wang, L.M. Fu, "Micropumps and biomedical applications – A review", *Microelectron. Eng.* 195:121–138, 2018. ISSN 0167-9317, https://doi.org/10.1016/j.mee.2018.04.008
15. N.T Nguyen, X. Huang, T.K. Chuan, "MEMS-micropumps: A review", *J. Fluids Eng.* 124:384–392, 2012.
16. D.J. Laser, J.G. Santiago, "A review of micropumps", *J. Micromech. Microeng.* 14:R35–R64, 2004.
17. J.D. Zahn, W. Wang, W. Soper, "Micropump applications in bio-MEMS", *Bio-MEMS–Technol. Appl.* 4:142–176, 2007.
18. M.W. Ashraf, S. Tayabba, N. Afzulpurkar, "Micro electromechanical systems (MEMS) based microfluidic devices for biomedical applications," *Int. J. Mol. Sci.* 12:3648–3704, 2011.
19. A.K. Yetisen, M.S. Akram, C.R. Lowe, "Paper-based microfluidic point-of-care diagnostic devices", *Lab. Chip.* 13:2210–2251, 2013.

20. G.S. Jeong, J. Oh, S.B. Kim, M.R. Dokmeci, H. Bae, S.H. Lee, A. Khademhosseini, "Siphon-driven microfluidic passive pump with a yarn flow resistance controller". *Lab. Chip.* 14:4213–4219, 2014.

21. Holzer, I.L. Haifa, D. Daniel, I.L. RaAnana, E. Hirszowicz, I.L. Ramat-Gan, "Methods for Treatment of Bladder Cancer", US Patent US9011411B2, 2004.

22. P. Enggaard, D.K. Vejby, C.S. Moller, D.K. Fredensborg, T. HedeMarkussen, D.K. Bagsvaerd, "Dose Mechanism for an Injection Device for Limiting a Dose Setting Corresponding to the Amount of Medicament Left", US Patent US9775953, 2017.

23. V.V. Yuzhakov, "The admin pen microneedle device for painless & convenient drug delivery", *Drug Deliv. Technol.* 10(4):32–36, 2010.

24. T. Tanner, R. Marks, "Delivering drugs by the transdermal route: Review and comment," *Skin Res. Technol.* 14(3):249–260, 2008.

25. F.K. Forster, R.L. Bardell, M.A. Afromowitz, N.R. Sharma, A. Blanchard, "Design, Fabrication and testing of fixed value micropumps", *Proceedings of the ASME Fluids Engineerinng Division, ASME*, 234, 1995.

9 Piezoelectric Actuator-based Micropumps

9.1 INTRODUCTION

A microelectromechanical system is a rapidly growing field which enables the manufacture of small devices using microfabrication techniques similar to the ones that are used to create integrated circuits. MEMS technologies have been applied to the needs of the biomedical industry giving rise to a new emerging field called microfluidics. Microfluidics deals with the design and development of miniature devices which can sense, pump, mix, monitor and control small volumes of fluids. The development of microfluidic systems has rapidly expanded to a wide variety of fields. Principal applications of microfluidic systems are for biological and chemical sensing, drug delivery system (DDS), lab-on-a-chip (LOC), point-of-care testing (POCT), molecular separation such as DNA analysis, amplification and for environmental monitoring. We saw in the last chapter that there are different micropumps controlling fluid flow in the microdomain. Their choice depends upon the application they are intended for. For drug delivery applications, piezoelectric micropumps emerged as a quite promising field in the last five decades.

Piezoelectric micropumps consist of a pumping chamber connected to two valves and a piezo-actuated membrane. The application of an alternating voltage across the piezo material (usually PZT) in the transverse direction causes the radial deformation and axial deflection of the membrane, thus changing the volume of the pumping chamber. These micropumps generate a large amount of force; however, a high-voltage power supply is required for their actuation. They can achieve moderate flow rates with high back pressures (if valves are suitably designed). However, they can consume much power if high actuation frequency and large discs are employed. The popular piezoelectric materials are PZT, lithium niobate ($LiNbO_3$), zinc oxide (ZnO) and BZT-BCT (Pb free piezoelectric), which are ceramic, and polyvinylidene fluoride (PVDF), which is a polymer material.

If we try summarizing piezoelectric (mainly employing PZT actuator) micropumps research in the last 15 years [1–10], it can be stated that PZT actuator-based micropumps operate in voltage ranges which are generally higher than 50 V. This range is above a human safe limit. Various thicknesses of PZT actuators have been employed, like 150–200 μm. Different kinds of design modifications were adopted by research groups to achieve high deflections, like foldable actuators and composite PZT materials [11,12]. Figures 9.1a–c summarize the trend of piezoelectric micropump technology based on input voltages used for piezoelectric actuation, maximum membrane deflection obtained, and thickness of piezoelectric actuator used.

DOI: 10.1201/9781003202264-9

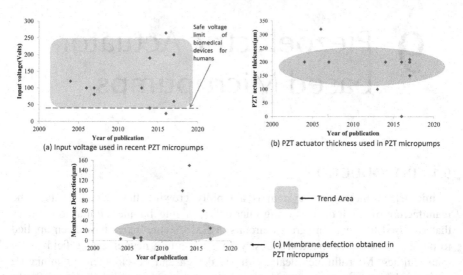

FIGURE 9.1 (a) Excitation voltages used in recent PZT microumps, (b) PZT actuator thickness used in PZT micropumps and (c) membrane deflection obtained in PZT micropumps.

However, piezoelectric actuators have the disadvantage of being heavy in weight, expensive in manufacturing and are less compliant to be safely operated near humans [13,14].

In the subsequent sections in this chapter, we shall look into the piezoelectric actuation in detail, carry on the theoretical analysis for a piezoelectric actuator for a micropump and then use a finite element-based method to look at micropump actuation technique. Finally, an approach for fabricating piezoelectric micropump using photolithography and soft micromoulding technique is presented.

9.2 PIEZOELECTRIC EXCITATION

Piezoelectric force has been widely used for micromechanical devices. If special crystals were subject to mechanical tension, they became electrically polarized and the polarization was proportional to the extension. They also discovered that the opposite was true if an electrical field was applied across the material it deformed. This is known as the inverse piezoelectric effect. Piezoelectricity involves the interaction between the electrical and mechanical behaviour of the medium. To the first order this is described as [15]

$$S = s^E T + dE \tag{9.1}$$

where S is the strain, s^E is the compliance tensor under conditions of constant electric field, T is the stress, d is the piezoelectric charge constant tensor and E is the electric field. The deformation of a piezoelectric crystal is illustrated in Figure 9.2. In the absence of mechanical loads Eq. (9.1) gives

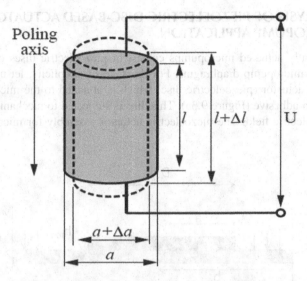

FIGURE 9.2   The deformation of a piezoelectric device when subject to an electrical voltage.

$$\Delta l = d_{33}.U = d_{33}.\frac{U}{l}.l = d_{33}.E.l \qquad (9.2)$$

and

$$\Delta l = d_{31}.U = d_{31}.\frac{U}{l}.a = d_{31}.E.a \qquad (9.3)$$

where Δl is elongation along the poling axis, l is the device length along the poling axis, U is the electrical voltage, Δa is elongation perpendicular the poling axis and a is the device length perpendicular to the poling axis. Normally $d_{33} > 0$ and $d_{31} < 0$.

Examples of piezoelectric materials are quartz, $LiTaO_3$, PZT and ZnO. Non-piezoelectric materials, e.g. silicon, can be excited by a piezoelectric material, e.g. PZT or ZnO. Another solution is to mount a piezoelectric disc on the non-piezoelectric material. This eliminates the problem of making the film thick enough for high voltages to be applied without dielectric breakdown (sparks/short circuits across the film). The piezoelectric effect can be used to bend a diaphragm, e.g., in a pump. When a voltage is applied across the piezoelectric disc it deforms and forces the diaphragm to bend.

The piezoelectric actuation has an added advantage in power consumption because it draws current only during the transition period. Two main uses of a piezoelectric actuator in microvalve applications are reported in the literature: one is taking advantage of the displacement parallel to the polarization (d33), while the other is utilizing the strain orthogonal to the polarization orientation (d31). Piezoelectric actuation has good reliability also better actuation force and response time than other actuation types.

9.3 ANALYSIS OF PIEZOELECTRIC DISC-BASED ACTUATOR FOR MICROPUMP APPLICATION

Piezoelectrically actuated micropumps consist of piezoelectric discs or piezoelectric films on micropump diaphragms. For the sake of simplicity, let us consider a piezoelectric actuator (piezoelectric disc) which is attached to the micropump diaphragm by an adhesive (Figure 9.3a). This disc is subjected to mechanical strain by an external electric field. The piezoelectric actuator assembly for micropumps can

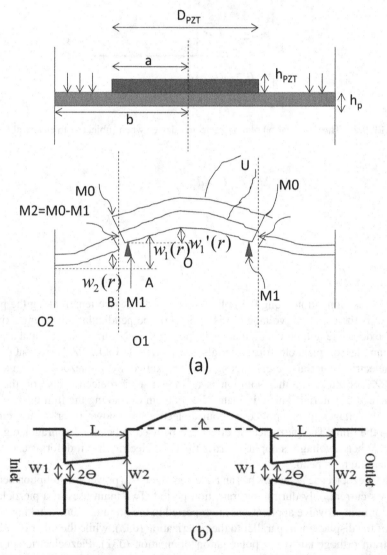

FIGURE 9.3   (a) Side view of piezoelectric actuator and membrane and (b) side view of the micropump chamber in supply mode.

TABLE 9.1

Parameter Values for the Theoretical Analysis of Micropump

Symbol	Parameter	Symbol	Parameter
D_e	Equivalent Flexural Modulus	Dp	Flexural Modulus of Passive Plate
a	Radius of PZT disc	b	Radius of whole membrane plate
v_e	Equivalent Posson's ratio	v_p	Poisson's ratio of passive plate
M_2	Intermediate moment applied at the outer edge of the two-layer structure	M_1	Intermediate moment applied at the inner edge of the ring structure

be thought to be made of three layers: PZT layer, bonding layer and passive plate. When applying an alternating electrical field across the PZT layer, it will generate a reciprocating deflection in the direction vertical to the surface of the actuator. This deflection is then transferred to the pumping effect that drives the fluid inside the pump chamber flowing through the inlet/outlet. The total deflection consists of two parts: one is proportional to pressure difference exerted on the whole structure and another is the voltage applied across the PZT disc [16]. For a disc type bending actuator, the system can be modelled into two parts: one considering the sandwiched layer with piezostack area and other is outer PDMS ring as shown in Figure 9.3a–b. The assumptions made are that the whole structure is circumferentially symmetrical, the thickness of each layer is much smaller than its radius and the deformation is very small in comparison with the dimension of its structure, the thickness of the bonding layer is small in comparison with the thickness of the PZT disc and passive plate and therefore the influence of the bonding layer on the deflection of the whole structure is neglected, bonding between the PZT disc and passive plate is perfect and that the outer edge of the passive plate is fixed [16].

The whole pump structure is divided into two parts and therefore the deflections can be calculated separately. One is the passive plate with a ring-like structure and the other is the central part which is a two-layer structure (bonding layer has been neglected). The parameters considered for theoretical analysis of micropump are given in Table 9.1.

The deflection of the two-layer structure (central part) due to application of voltage is

$$w_1(r) = \frac{M_0\left[\left(b^2-a^2\right)\left(a^2-r^2\right)+a^2\left(a^2-2b^2\log\frac{a}{b}-b^2\right)\right]}{2\left\{D_p\left[\left(1+v_p\right)a^2+\left(1-v_p\right)b^2\right]+D_e\left(1+v_e\right)\left(b^2-a^2\right)\right\}} \qquad 0 \leq r \leq a \quad (9.4)$$

where r is the horizontal distance from the centre of the disc and M_0 is the intermediate moment caused by the actuation of PZT and is given by

$$M_0 = D_e \frac{-d_{31} U / h_{PZT}}{\frac{h}{2} + \frac{2}{h}\left(\frac{1}{E_{PZT} h_{PZT}} + \frac{1}{E_p h_p}\right)(D_{PZT} + D_p)} \tag{9.5}$$

where the equivalent parameters of the composite structure are given by

$$E_e = C_1 E_{PZT} + C_2 E_P + \frac{C_1 C_2 E_{PZT} E_P (\vartheta_{PZT} - \vartheta_p)^2}{C_1 E_{PZT}\left(1 - \vartheta_P^2\right) + C_2 E_P\left(1 - \vartheta_{PZT}^2\right)} \tag{9.6}$$

$$\vartheta_e = \frac{C_1 \vartheta_{PZT} E_{PZT}\left(1 - \vartheta_P^2\right) + C_2 \vartheta_P E_P\left(1 - \vartheta_{PZT}^2\right)}{C_1 E_{PZT}\left(1 - \vartheta_P^2\right) + C_2 E_P\left(1 - \vartheta_{PZT}^2\right)} \tag{9.7}$$

$$D_e = \frac{E_e h^3}{12\left(1 - \vartheta_e^2\right)} \tag{9.8}$$

$$C_1 = \frac{h_{PZT}}{h} \tag{9.9}$$

$$C_2 = \frac{h_P}{h} \tag{9.10}$$

The deflection of the passive membrane ring (outer part) is

$$w_2(r) = \frac{M_0\left[a^2\left(r^2 - 2b^2 \log\frac{r}{b} - b^2\right)\right]}{2\left\{D_p\left[\left(1 + \upsilon_p\right)a^2 + \left(1 - \upsilon_p\right)b^2\right] + D_e\left(1 + \upsilon_e\right)\left(b^2 - a^2\right)\right\}} \quad a \le r \le b \tag{9.11}$$

Maximum deflection occurs at centre i.e. at r=0. So

$$w_{\max}(r) = \frac{M_0\left[\left(b^2 - a^2\right)\left(a^2\right) + a^2\left(a^2 - 2b^2 \log\frac{a}{b} - b^2\right)\right]}{2\left\{D_p\left[\left(1 + \upsilon_p\right)a^2 + \left(1 - \upsilon_p\right)b^2\right] + D_e\left(1 + \upsilon_e\right)\left(b^2 - a^2\right)\right\}} \tag{9.12}$$

The equivalent material properties of the two-layer structure can also be calculated as per above work. The deflection caused by mechanical pressure difference is also calculated respectively for the two sections: the central part and the outer part. Under constant pressure difference p, the deflection of the two-layer structure is

$$w_3(r) = \frac{p}{64D_p}(b^2-a^2)^2 \frac{(M_1-M_2)a^2\left(a^2-2b^2\log\frac{a}{b}-b^2\right)}{2D_p\left[(1+\upsilon_p)a^2+(1-\upsilon_p)b^2\right]}$$
$$+\frac{p}{64D_e}\left(\frac{5+\upsilon_e}{1+\upsilon_e}a^2-r^2\right)+\frac{M_2(a^2-r^2)}{2D_e(1+\upsilon_e)} \quad 0\le r\le a$$

(9.13)

and the deflection of the passive ring is

$$w_4(r) = \frac{p}{64D_p}(b^2-r^2)^2 \frac{(M_1-M_2)a^2\left(r^2-2b^2\log\frac{r}{b}-b^2\right)}{2D_p\left[(1+\upsilon_p)a^2+(1-\upsilon_p)b^2\right]} \quad a\le r\le b \quad (9.14)$$

In order to understand more clearly, we can consider by taking a set of parameters for our analysis. The piezoelectric disc material is PZT-5H and is glued on polydimethyl siloxane (PDMS) membrane, as shown in Figure 9.3a (radius of PZT disc (a) – 4 mm, radius of PDMS membrane (b) – 4.5 mm, Thickness of PZT Disc (h_{PZT}) = 1 mm, Thickness of PDMS membrane(hp) = 0.25 mm, Young's Modulus of PZT Disc (E_{PZT}) = 67 GPa, Young's Modulus of PDMS Membrane (E_p) = 180 kPa, Poisson's ratio of PZTdisc (σ_{PZT}) = 0.31, Poisson's ratio of PDMS membrane (σ_p) = 0.5, Piezoelectric Charge Coefficient (d_{31} =−1.9e-10 m/V). Here, we neglect the deflection due to pressure (since its magnitude is very less). Then, the maximum deflection of the centre point of the membrane comes to be around 0.02 μm. This deflection changes with different parameters of piezoelectric disc, membrane and applied voltage. In the above equations, damping is neglected. The deflection of the membrane due to piezo-electric actuation will be dampened if air, water or other liquid drug formulations are considered. Hence medium damping should be considered from the very beginning of the design stage of a micropump.

9.4 CASE STUDY – PIEZOELECTRIC ACTUATOR-BASED VALVELESS NOZZLE DIFFUSER STRUCTURE MICROPUMP

In the earlier chapters, we have seen that a micropump may have a check-valve-based structure or valveless nozzle-diffuser-based structure. For a single chamber check-valve structure, the actuator alternatively increases and decreases the pump chamber volume by applying force to the diaphragm. Fluid is drawn in during the expansion (suction) stroke and forced out during the contraction (discharge) stroke. Check valves are used for orienting the flow into and out of the pump chamber. Figures of merit of this structure are critical opening pressure, ratio between forward and reverse pressure drop, ease of fabrication and reliability. But valve fabrication is a difficult task as compared to valveless structures. Whereas we observe for a single chamber valveless structure that the structure is similar to check-valve pumps. The only difference is the use of diffusers/nozzles instead of check valves. The flow channels at the inlet and the outlet are designed to give different flow resistance in the forward and reverse directions. With correctly designed diffuser elements, more fluid flows through the inlet element than through the outlet element during the supply mode. During the pump mode, more fluid flows through the outlet element than through the inlet element. This results in a net flow from the inlet side to the outlet side of the pump. Benefits of this structure are relatively simple construction in comparison to pump concepts with check valves, pumping of particle-loaded media, such as cells, is easier to achieve due to the open flow structures. Drawbacks of the structure are that no self-blocking effect exists; any over pressure build-up at the outlet will cause reverse flow that becomes predominant as soon as the pump is switched off.

9.4.1 OPERATING PRINCIPLE OF VALVELESS MICROPUMP

The diffuser/nozzle element is the key part of the serial valveless micropump. The flow rectification of the valveless micropump depends on the difference of pressure loss in diffuser direction and nozzle direction. The pressure loss coefficient through diffuser/nozzle elements is defined as the ratio of pressure head drop to the velocity head. The valveless diffuser micropump works in two modes: supply mode and pump mode. In the supply mode chamber volume increases. Nozzle/diffuser elements work in such a way that the fluid flow from the inlet to the pump chamber dominates due to diffuser action from the inlet to the chamber and nozzle action from the outlet to the chamber, i.e. more fluid enters into the chamber through the inlet than fluid through the outlet. While in pump mode the chamber volume decreases. The fluid flow from the pump chamber to the outlet dominates in this mode rather than the flow from the chamber to the inlet due to the diffuser action from the chamber to the outlet and the nozzle action from the chamber to the inlet. The two modes (supply and pump) are shown in Figure 9.4a.

Piezoelectric actuation is the most widely used for actuating diaphragm micropumps achieving relatively large displacement magnitudes and forces. But the piezoelectric actuators have the disadvantage of being heavy in weight, expensive in manufacturing and less compliant to be safely operated near humans [13,14]. In this example, some of the commercially available piezoelectric discs (PZT 5-H material) are considered to determine their resonant frequency and deflection with applied voltage. The micropump structure with its components that shall be used throughout this example is shown below.

9.4.2 SIMULATION OF PIEZOELECTRIC MICROPUMP

9.4.2.1 Micropump Components

Like earlier chapters, finite element analysis-based simulation is used to optimize the valveless nozzle diffuser structure of the micropump. Comsol Multiphysics 5.2 is used for the simulations for this example. Let us consider the initial dimensions of diffuser as inlet (W1) – 300 μm, outlet (W2) – 750 μm and length of diffuser (L) – 1700 μm) (shown in Figure 9.5a) [16]. The meshing of the diffuser structure

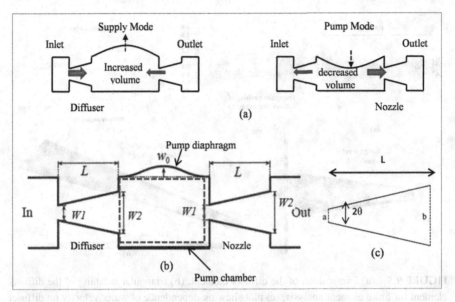

Supply Mode

Inlet   Outlet

Increased
volume

Diffuser

Pump Mode

Inlet   Outlet

decreased
volume

Nozzle

(a)

Pump diaphragm

W_0

L L

In $W1$ $W2$ $W1$ $W2$ Out

Diffuser   Nozzle

(b)

Pump chamber

L

a 2θ b

(c)

FIGURE 9.4   Operational modes, (a) supply mode and pump mode of a valveless piezoelectric micropump, (b) schematic of the micropump and (c) schematic of diffuser element.

for finite element method is shown in Figure 9.5b. Further it is assumed that pressure difference of 100 Pa is applied between input and output of diffuser. A flat-walled diffuser structure is considered as it gives the same performance as the conical diffuser but in a compact structure and is easier to fabricate [17]. The simulation studies shall explore the structural module as well the microfluidics module (laminar flow) of Comsol Multiphysics. In laminar flow module, the no-slip boundary wall condition is selected i.e. at walls velocity of fluid becomes zero. This condition is chosen as we are dealing with flow in the microfluidic domain and essentially laminar regime. For the diffuser structure, the losses are minimum when twice the diffuser angle is near to 5° but volume flowrate increases with 2θ (Figure 9.5c).

With a higher diffuser angle the flow gets stalled towards the boundaries. Considering this, a diffuser angle of 10° is chosen as optimum (Figure 9.5d). To observe the pressure variation along length of diffuser, pressure at outlet, P_{out} is fixed at 50 KPa and input water velocity of 0.1 m/s is given. It is seen that there is a variation of pressure difference of around 220 Pa from the inlet to the outlet. Velocity at the boundary walls is zero and it is maximum along the centre of the diffuser. Velocity decreases along the length of the diffuser except for a sudden rise in velocity near the input of diffuser (at 50 μm from input). The volume flow rate of the diffuser varies 18.5 μl/min at the inlet to around 24 μl/min at the outlet of the diffuser.

Next, the focus shall be the design of the micropump structure and simulation of velocity and pressure through the micropump chamber. The diameter of the micro pump chamber is around 10 mm. The nozzle and diffuser dimensions remain the same as considered above. The fluid considered is water. A pressure difference of 1000 Pa is applied between the input and output of the micropump. A triangular mesh

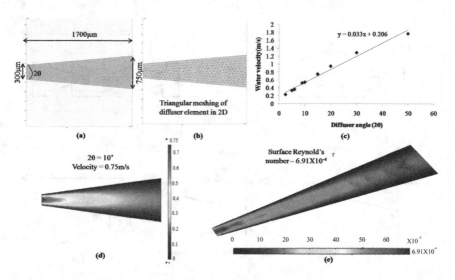

FIGURE 9.5 (a) Dimensions of the diffuser element, (b) triangular meshing of the diffuser element for finite element analysis, (c) plot showing dependence of water velocity on diffuser angle, (d) velocity distribution through the nozzle structure where velocity is maximum at the inlet and (e) variation of Reynold's Number.

is considered for meshing (Figure 9.6a). A variation of pressure along the micropump structure is shown (Figure 9.6b) where the inlet pressure is the highest, decreases in the diffuser, remains constant throughout the chamber and drops further down to zero at the output of the nozzle. The variation of velocity in the micropump structure is shown in Figure 9.6c where velocity is high (around 0.85 m/s) near the diffuser inlet, somewhat decreases throughout the chamber length to around 0.4 m/s and then rapidly increases to 1.1 m/s at the nozzle inlet. Thus the flat walled nozzle diffuser pump is able to produce a net increase in volume flowrate. Figure 9.6d shows the Reynold's Number variation in the micropump chamber which comes to 42.2 indicating a laminar flow. Based upon the simulation results, the micropump components dimensions are summarized in Table 9.2.

9.4.2.2 Piezoelectric Actuator Simulation

The structural and piezoelectric module of Comsol Multiphysics can be used for piezoelectric disc and actuator assembly simulation. The parameters considered for simulation are membrane radius – 4500 μm, membrane thickness – 80 μm, PZT-5H disc radius – 4000 μm and PZT-5H disc thickness – 150 μm, applied voltage 60 V. Figure 9.7 shows the simulations results.

As expected, the deflection of the pump membrane increases with increasing voltage (Figure 9.7a) and varies with different kinds of piezoelectric disc material chosen. PZT 5H material gives maximum deflection among other piezoelectric materials like PZT 5 A, PZT–8, barium titanate and zinc oxide. Figure 9.7b shows the deflection of the centre point of the membrane when piezoelectric radii are increased on keeping the membrane radii fixed at different values. It is seen that a ratio of 0.88 between the

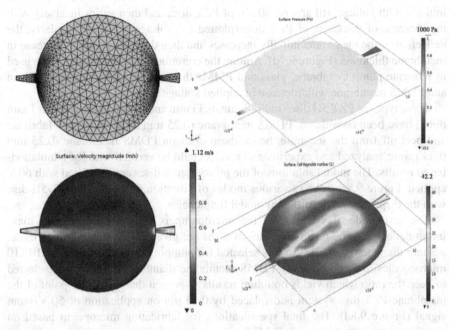

FIGURE 9.6 (a) Triangular meshing of micropump chamber, (b) pressure variation across micropump chamber and (c) velocity variation across micropump chamber and Reynold's Number variation across micropump chamber.

TABLE 9.2
Micropump Components Dimensions Considered after Simulation Studies

S/n	Parameter	Value
1	Chamber diameter	9000 μm
2	Micropump chamber depth	100 μm
3	Diffuser length	1700 μm
4	Diffuser inlet width	300 μm
5	Diffuser outlet width	750 μm
6	Divergence angle	10°

piezoelectric disc radius and the membrane radius gives the optimum performance [16]. For a given radius of the pump membrane (5000 μm, 4000 μm and 2500 μm in our case), the deflection of the membrane initially increases and then decreases with the radius of the piezoelectric disc (Figure 9.7b). It is seen that maximum defection of pump membrane is achieved at the piezoelectric disc radius and membrane radius ratio of 0.8. That is if the piezoelectric disc radius is 4000 μm then membrane radius should be 4545 μm. Figure 9.7c shows that the displacement of membrane increases initially with voltage, till around 60 μm of PZT disc, and then varies inversely with the increase of thickness of PZT disc (plotted for voltage = 60 V). Similarly, the deflection of the membrane initially increases and then decreases with an increase in membrane thickness (Figure 9.7d). Among the common materials which can be used to fabricate pump membrane, glass and PMMA have higher deflection than silicon and PDMS membrane with increase in applied voltage.

Three types of PZT 5H discs radii (8 mm, 6.35 mm and 5 mm diameter, each 1 mm thick) have been considered. PDMS membrane (0.25 mm thick) is easy to fabricate and peel off from the substrate, hence these discs on PDMS membrane (0.25 mm thick) were analyzed and experimental results could be verified with the simulation-based results. The modal solutions of the piezoelectric discs are simulated with 60 V applied. Figure 9.8a–c shows various modes of vibration of the 8 mm PZT-5H disc with the displacements at different modal frequencies.

The first mode of vibration of the micropump is chosen since it gives maximum deflection. The results for various discs are given in Table 9.3. Out of these, a 8 mm diameter PZT 5H disc was selected for pumping drug to an array of 10X10 microneedles and consequently PDMS membrane diameter of 9 mm was selected to keep the ratio which yields optimum results. It is seen that the centre point of the membrane-PZT disc system is displaced by 0.13 μm on application of 50 V input signal (Figure 9.8d). The final specifications for fabricating micropump based on simulation results are shown in Table 9.4.

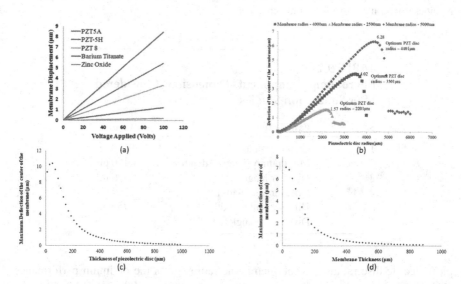

FIGURE 9.7   (a) Variation of deflection of centre of the membrane with applied voltage for different piezoelectric materials, and (b) variation of deflection of centre of the membrane with piezoelectric disc radius for fixed membrane radius. (c) Variation of deflection of centre of the membrane with thickness of the piezoelectric disc keeping other parameters fixed, and (d) variation of deflection of centre of the membrane with membrane thickness keeping other parameters fixed.

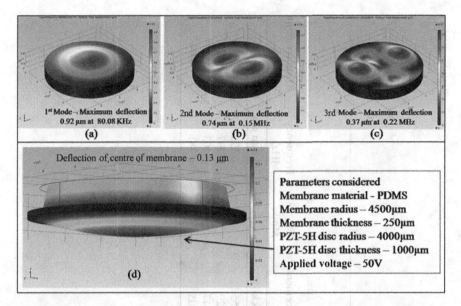

FIGURE 9.8 (a) Modal solutions for PZT disc (8 mm diameter and 1 mm thickness). There are four modes of vibration. (b) Deflection of PZT-5H disc attached to PDMS membrane and 50 V applied.

It is seen that the centre point of the membrane-PZT disc system is displaced by 0.13 μm on application of 50 V input signal (Figure 9.8d). The final specifications for fabricating micropump based on simulation results are shown in Table 9.4.

9.4.3 Fabrication Approach

The following steps may be adapted for fabrication of a piezoelectric micropump. The micropump chamber is fabricated using polydimethylsiloxane (PDMS) and then bound on a glass slide after surface modification. PDMS has several properties like biocompatibility, low cost and transparency which make it favourable for use in creating microfluidic devices. PZT 5H material disc (8 mm wide, 1 mm thickness) is used as a piezoelectric actuator. These steps are also illustrated using the schematic in Figure 9.9.

(a) Silicon wafer with silicon dioxide layer grown on top of it is chosen as substrate.
(b) SU-8 2050 is spun on glass substrate yielding around 50 μm film. It is placed on a hotplate at 65°C and 90°C, depending upon the datasheet specification for the coated thickness (process known as softbake).
(c) SU-8 is exposed to UV light through mask. The timing of exposure is calculated as per the dose requirements. This leads to crosslinking of SU-8 polymer in the places where light reaches the photoresist coating. It is again

TABLE 9.3
Modal Solutions for Different PZT-5H Discs Considered for Simulation

Mode	Frequency (MHz)	Displacement (µm)	Mode	Frequency (MHz)	Displacement (µm)	Mode	Frequency (MHz)	Displacement (µm)
Disc A (8 mm dia, 1 mm thick)			Disc B (6.35 mm dia, 1 mm thick)			Disc C (5 mm dia, 1 mm thick)		
1	0.08	0.92	1	0.118	0.55	1	0.1738	0.33
2	0.14	0.45	2	0.2108	0.29	2	0.297	0.2
3	0.15	0.74	3	0.2163	0.47	3	0.303	0.27
4	0.37	0.37	4	0.2962	0.05	4	0.381	0.04

TABLE 9.4
Optimized Values for Micropump after Simulation

S/n	Parameter	Optimized Value from Simulation	Chosen Value	Reason for Choice
1	Piezoelectric material	PZT-5H	PZT-5H	Maximum deflection among piezoelectric materials
2	PZT disc diameter	8 mm	8 mm	Commercially commonly available PZT disc parameter
3	PZT disc thickness	60μm	1 mm	Commercially commonly available PZT disc parameter
4	Membrane material	PDMS	PDMS	Biocompatible, ease of fabrication by soft moulding
5	Resonant frequency (8 mm disc)	80 KHz	80 KHz	
6	Resonance mode	1st	1st	Maximum deflection is obtained for the first mode
5	Membrane diameter	9 mm	9 mm	Optimum performance for given PZT-5H disc dimensions
6	Optimum membrane thickness	20–30 μm	250 μm	Ease of peeling membrane layer from substrate

     placed on a hotplate at 65°C and 90°C for thickness specific timing (process known as hardbake). Sample then cooled gradually.

(d) Development of SU-8 in SU-8 developer to remove the uncrosslinked SU-8 thus yielding master mould pattern having inverse pattern for a micropump. This mould can now be used multiple times for soft lithography.

(e) For creating the micropump structure, PDMS and curing agent mixture (ratio 1:1) is freed from air bubbles using a vacuum pump and then spin coated on the SU-8 at speed governed by desired thickness. The PDMS mixture is then cured at 70°C for 30 minutes.

(f) After curing, PDMS layer is peeled gently to release it from the wafer and SU-8 structure. If inlet and outlet holes are to be defined, then it may be done at this stage by using a punch.

(g) A glass disc is taken and the sides of the PDMS structure and glass slides which are to be joined together are kept in a plate and exposed to oxygen plasma for their surface modification / activation. On both substrates, the treatment is effective at removing hydrocarbon groups (CxHy) leaving behind silanol groups on the PDMS and OH groups on the glass substrate respectively. This allows strong Si–O–Si covalent bonds to form between the two materials.

(h) Withing 2–3 minutes of surface modification, the PDMS structure is gently placed on the glass slide removing any trapped air bubble in the process. This strongly bonds the PDMS micropump structure on the glass substrate.

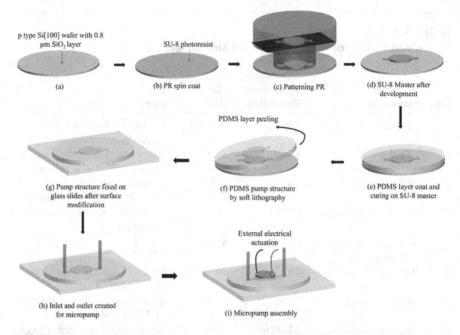

FIGURE 9.9   Schematic showing piezoelectric micropump fabrication steps.

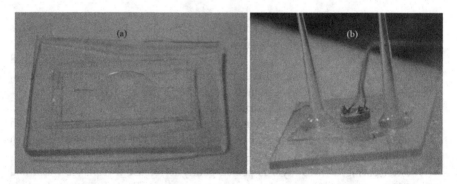

FIGURE 9.10   (a) PDMS micropump structure fabricated by soft lithography process and bonded on glass after surface plasma modification and (b) micropump with inlet and outlet and PZT-5H disc adhered on the micropump structure.

(i)   The piezoelectric disc is fixed on top of the chamber with help of adhesive and electrical connections soldered for external electrical actuation. Inlet and outlet tubes are also fixed using adhesive. O make leakage free structures, strong adhesives are recommended.

The final fabricated structure resulting after following above steps is shown in Figure 9.10.

9.5 CHARACTERIZATION OF PIEZOELECTRIC ACTUATOR

The piezoelectric actuator can be characterized for its various properties using an equipment known as a Laser Doppler Vibrometer. It makes non-contact vibration measurements of a surface. The laser beam is directed on the sample and the frequency and amplitude of vibration are measured from the Doppler shift observed in the reflected laser beam as picked up by the detector. For the presented case study, Laser Doppler Vibrometery (Polytec, Germany) was used and is shown in Figure 9.11.

Figure 9.12a shows the resonant frequency obtained for the PZT disc (8 mm diameter, 1 mm thickness) in air when 8 V sinusoidal signal is applied. This resonant frequency (297.42 KHz) might differ from the simulated value which one obtains during simulation. There are several reasons for it. Our example neglected the damping effects, and the deflection was calculated considering vacuum. This led to variation in the simulated value and the characterized values. Hence it is recommended that

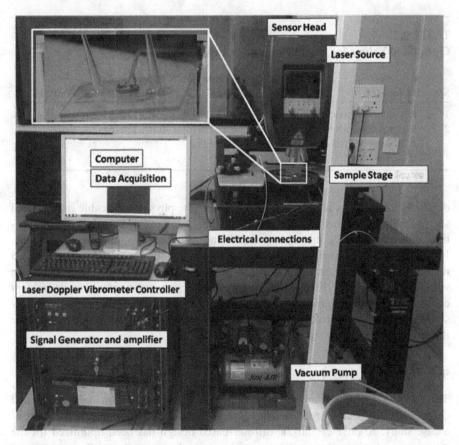

FIGURE 9.11 Laser Doppler Vibrometer-based setup for characterization of piezoelectric actuator, courtesy Microelectronics and MEMS laboratory, Department of Electronics and Electrical Communication Engineering, Indian Institute of Technology, Kharagpur, India.

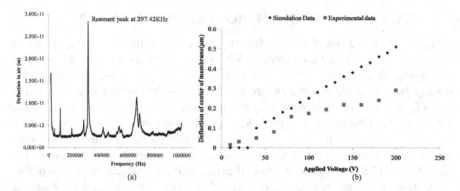

(a)   (b)

FIGURE 9.12   (a) Experimentally determined resonant frequency of 8 mm diameter PZT disc at 8 V input signal and (b) comparison of simulation and experimental data for 8 mm diameter PZT disc's deflection with varying voltage.

TABLE 9.5
Optimized Values for Micropump after Fabrication

S/n	Parameter	Simulation Results	Experimental Results	Remarks
1	Resonant frequency (8 mm disc)	80 KHz	297.42 KHz	
2	Maximum displacement of centre of 8 mm PZT disc at 10 V	0.12 μm in vacuum	0.092 μm in air	Variation due to damping due to air

during simulation one may try to simulate using the physics applicable in the actual scenario. Figure 9.12b shows the plot of simulated and experimental data for deflection of the centre of the micropump membrane with increasing input signal voltage. It is noted that the experimental values are lower than the simulated values for each reading. This may be due the fact that simulated values considered deflection in vacuum while the experimentally determined deflection was measured in air ambient conditions. This may have introduced damping in the deflections. Table 9.5 below summarizes the comparison of simulated and experimentally determined values for the micropump as considered in the example.

9.6   CONCLUSION

This chapter presented an overview of the analysis, simulation, fabrication and characterization studies required for a piezoelectric micropump with the help of a case study. A valveless nozzle-diffuser-based design has been optimized for the micropump. A piezoelectric micropump having valveless nozzle diffuser structure, PDMS membrane and PZT-5H actuator was successfully fabricated. The simulated resonant frequency of PZT discs differed from experimentally determined values because of non-consideration of effects of air damping in simulation. The simulated and experimentally determined values for PZT disc deflection were of same order The experimentally determined values were slightly less due to air damping. Overall, the deflections obtained by commercially available PZT disc is of submicron value even at high (>100) voltages. Biocompatibility of lead-based PZT micropumps are an issue due to lead related poisoning. Moreover, different government agencies like the US Food and Drug Administration (FDA), Center for Disease and Control & Prevention (CDC), and Restriction of Hazardous Substances Directive (RoHS) have issued negative findings against lead-related materials and devices, implying unsuitability of PZT micropumps for drug delivery applications. Hence, the following chapters deal with the ionic polymer metal composite-based micropumps which achieve good deflection at relatively low voltages (3–8 V).

REFERENCES

1. W.W. Chi, I. Hj Abdul Azid, "Comparison of the performances of micropump with active type diaphragm actuated by several approaches," *2008 33rd IEEE/CPMT International Electronics Manufacturing Technology Conference (IEMT)*, Penang, 1–4, 2008.

2. D.S. Lee, J.S. Ko, Y.T. Kim, "Bidirectional pumping properties of a peristaltic piezoelectric micropump with simple design and chemical resistance," *Thin Solid Films.* 486:285–290, 2004.

3. T. Zhang, Q.M. Wang, "Performance evaluation of a valveless micropump driven by a ring-type piezoelectric actuator", *IEEE Trans. Ultrason. Ferroelectr. Freq. Control* 53(2), 463–73, 2006.

4. L.S. Jang, W.H. Kan, "Peristaltic piezoelectric micropump system for biomedical applications," *Biomed. Microdevices* 9(4):619–626, 2007.

5. A. Geipel, A. Doll, P. Jantscheff, N. Esser, U. Massing, P. Woias, F. Goldschmidtboeing, "A novel two-stage backpressure independent micropump: Modeling and characterization", *J. Micromech. Microeng.* 17:949–959, 2007.

6. C. Hernandeza, Y. Bernard, A. Razek, "A global assessment of piezoelectric actuated micro-pumps", *Eur. Phys. J. Appl. Phys.* 51(20101), 2010. DOI: 10.1051/epjap/2010090

7. F.R. Munas, G. Melroy, C. Bhagya Abeynayake, H.L. Chathuranga, R. Amarasinghe, P. Kumarage[1], V.T. Dau, D.V. Dao, "Development of PZT actuated valveless micropump, *Sensors*, 18(5):1302, 2018.

8. T. Wang, J. He, C. An, J. Wang, L. Lv, W. Luo, C. Wu, Y. Shuai, W. Zhang, C. Lee, "Study of the vortex based virtual valve micropump", *J. Micromech. Microeng.* 28(125007):8, 2018.

9. H.K. Ma, B.R. Hou, H.Y. Wu, C.Y. Lin, J.J. Gao, M.C. Kou, Development and application of a diaphragm micropump with piezoelectric device, *In Symp. Des. Test Integr. Packag. MEMS/MOEMS, DTIP, Stresa*, Italy: Lago Maggiore, 273–278, 2007.

10. B. Pečar, D. Križaj, D. Vrtačnik, D. Resnik, T.D.M. Možek, "Piezoelectric peristaltic micropump witha single actuator", *J. Micromech. Microeng.* 24(105010):9, 2014.

11. X.Y. Wang, Y.T. Ma, G.Y. Yan, Z.H. Feng, "A compact and high flow-rate piezoelectric micropump with a folded vibrator", *Smart Mater. Struct.*, 23(115005), 11, 2014.

12. Y. Li, P. Zhang, "Study on mechanical properties for modeling and simulation of microneedles for medical applications", *Appl. Mech. Mater.* 454:86–89, 2013.

13. L. Hines, K. Petersen, G.Z. Lum, M. Sitti, "Soft actuators for small-scale robotics", *Advan. Mater.* 29(1603483), 2017.
14. D. Rus, M.T. Tolley, "Design, fabrication and control of soft robots", *Nature* 521(7553):467, 2015.
15. S. Das, *"A Review on Micropump"*, M.Tech thesis, India: Institute of Radiophysics and Electronics, 2004.
16. B. Pramanick, P.K. Dey, S. Das, T.K. Bhattacharyya, "PDMS Membrane based SU-8 Micropump for drug delivery system", *J. ISSS.* 2, 2013.
17. A. Olsson, G. Stemme, E. Stemme, "Numerical and experimental studies of flat-walled diffuser elements for valve-less micropumps", *Sens. Actuator A: Phys.* 84(1–2):1, 165–175, 2000.

10 Ionic Polymer Metal Composite Actuators for Micropumps

10.1 INTRODUCTION

Over the years, different actuation techniques for micropumps have been explored to find a suitable material for micropump-based drug delivery application. Let us look at the information we have gathered about micropump actuation techniques discussions in the last few chapters.

- In the last chapter we saw that piezoelectrically actuated micro pumps usually produce high actuation forces and fast mechanical responses, but they need high input voltages. They are also expensive to manufacture and less compliant to be safely operated near humans [1,2].
- Thermo-pneumatically actuated micro pumps need low input voltages, generate high pump rates, and can be very compact, but high power consumption and long thermal time constants are the main disadvantages.
- Electrostatically actuated micro pumps have the merits of fast response time, microelectromechanical system (MEMS) compatibility, and low power consumption, but small actuator stroke, degradation of performance, and high input voltage are the main deterrents to using this diaphragm.
- Electromagnetically actuated diaphragms micropumps have a fast response time, but they are not well compatible with MEMS and require high power consumption.

This leads to exploration of smart membranes for pumps, which is becoming the enabling technology for a micropump. The current trend is towards the electrically driven smart polymeric or gel-based actuators which change shape or size in response to electrical stimulus. In this chapter we are going to look at an ionic polymer metal composite-based smart membrane and understand its response to an electrical actuation. We shall see the feasibility of it being used as a micropump actuator. Similar to the previous chapter, our approach shall be based upon brief theoretical analysis, simulation-based validation and finally fabrication and characterization.

10.2 IONIC POLYMER METAL COMPOSITES

The electroactive polymer (EAP) class of actuators covers shape memory-based polymers, liquid crystal polymers, ionic gels and ionic polymer metal composites shape memory polymers can deform their shapes under certain imposed condition and return back to their original form when the imposed condition is removed [3].

DOI: 10.1201/9781003202264-10

Liquid crystal polymers have relatively stiff rod like molecules that realign themselves in the presence of an electric field [4]. Ionic gels have mesogenic groups attached to siloxane or acrylate polymer main chains which deform in the presence of an electric field [5]. Ionic polymer metal composites are generally ion exchange membranes which are sandwiched between noble metals such as gold or platinum as electrodes [6]. They have the advantages of low actuation voltages, operability in aqueous environments, are light weight and have a large range of deflections [7,8]. Electrode coating on IPMCs requires being uniform, strongly adhering to the membrane and providing low electrical resistance [9]. Nafion (Teflon) backbone is one of the readily commercially available IPMCs [10,11]. It is available in both liquid as well as sheet form (when cast into thin membranes). In the membrane structure two distinct regions, namely a hydrophobic region of fluorocarbon backbone and hydrophilic sulfonate region, are formed within the membrane [12,13]. The instances of Nafion membrane with platinum-coated electrodes (both as electroplated platinum or platinum by impregnation-reduction technique) are widespread in literature [9,14,15].

There are two types of EAPs based on the activation mechanism. Electronic EAPs are driven by coulombic forces which include piezoelectric, electrostatic, and ferroelectric polymers. They require a large activation voltage but are well controlled and show a fast response. Ionic EAPs exhibit large bending displacement due to diffusion or migration of ions across the polymer membrane and they consist of two electrodes and an electrolyte. They produce large deformation even at very low voltages. Ionic EAPs are favourable for micropump actuation as they require very low activation voltage for producing large displacements at a fairly quick speed. In this chapter we shall look at the promising and emerging electroactive polymer-based actuation technique for micropumps. This actuation technique is based on using ionic polymer metal composite (IPMC) membranes as actuators for displacement micropumps. This new group of smart materials shows large deformations at comparatively low actuation voltages and fairly quick response. As a case study, a Nafion-based IPMC actuator is modelled using COMSOL Multiphysics v.4.3 software. In this chapter this actuator is further shown to be used as the micropump diaphragm for a circular valveless micropump and presents readers an elaborate discussion on the topic.

A comparative study of IPMC with other smart materials is listed in Table 10.1.

TABLE 10.1
Comparative Study of IPMC with Other Smart Materials

Actuator Type	Strain (%)	Response Time	Efficiency (%)
Piezoelectric	0.2	Fast	>80
1. Ceramic	1.7	Fast	>80
2. Single-crystal	0.1	Fast	N/A
3. Polymer			
Shape Memory Alloy	>5	Slow	<10
Magnetostrictive	0.2	Fast	<60
Electrostrictive polymer	4	Fast	N/A
IPMC	>40	Medium-Fast	>40

10.3 IPMC AS MICROPUMP ACTUATOR

In the last decade a new breed of polymer has emerged which responds to external electrical stimulation by displaying a significant shape or size displacement. These materials are known as ionic polymer metal composites. Due to their small electrical energy consumption, light weight and compliant properties, bio compatibility, ability to operate in air and aquatic media, insensitivity to magnetic fields and simple fabrication processes, this class of smart materials has proven to be important for micropump fabrication in controlled-drug delivery systems.

The electro-chemical-mechanical bending mechanism of the IPMC actuator is shown in Figure 10.1. When a voltage is applied across the Nafion electrodes, then two boundary layers, consisting of anions and cations are created. The anion layers are fixed while the cations are movable. The cations dissociate from their cluster regions and rapidly migrate towards the cathode in response to the electric field [16–20]. A bending moment is created when these hydrophilic cations drive the water molecules along with them, to the cathode. This results in accumulation of water molecules near the cathode and swelling of the membrane, thus causing deformation. After some time, water diffuses back to the membrane body inducing relaxation of the membrane.

There have been a lot of attempts to explain the fundamental mechanisms occurring in IPMCs with different models. Shahinpoor and Nemat Nessar presented a physics-based IPMC model [21,22]. Todokoro et al. presented a model for ionic polymer Nafion Platinum actuators [23]. Mallavarapu and Leo et al. [24] discussed the feedback control of the ionic polymer actuator bending. Newbury and Leo [25,26] presented an electrical model for ionic polymer transducers. During the period of 2000–2006, Nemat Nessar et al. presented the electromechanical and micromechanical actuator

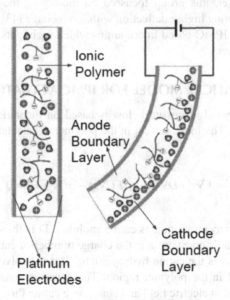

FIGURE 10.1 Electro-mechanical transduction of IPMC actuator.

model [27–29]. Lee et al. [30] presented the beam theory for IPMC modelling while Branco et al. [31] derived equivalent circuit representation of IPMC electromechanics. Wallmersperger and Leo developed a chemoelectromechanical model describing IPMC transduction [32–34]. He et al. [35] modelled the ion transport in IPMC using Monte Carlo simulation method.

The active use of Nafion-based micropumps can be traced to 2004 where Pak et al. used commercial Nafion 112 attached PDMS membrane on conventional valveless nozzle diffuser structure which deflected 14-21 μm and pumped at 9.97 μL/min at 4 V, 0.5 Hz [36]. Chung et al., in 2005, tested silver-plated Nafion membrane which gave 2.1 mm–6 mm deflection at 3 volts [37]. Simulation-based design modification in electrode design was done by Lee et al. where a ring-shaped electrode was patterned on circular Nafion membrane which deflected 0.9 mm at 2 V, 0.1 Hz with conical nozzle diffuser elements [38,39]. They also tried the square shaped IPMC diaphragm which gave 0.76 mm deflection for same signal. A flowrate of 492 μL/min was achieved with circular membrane design. Nguyen et al. developed Nafion/silica nanocomposite membrane-based micropump with PDMS check valves to achieve high deflection of 5.6 mm [40]. They could achieve an increased flowrate of 760 μL/min at 3 V input voltage and 3 Hz driving frequency with flap valve design.

Santos et al. used a circular Nafion polymer with smaller circle-shaped platinum electrode and four platinum bands and operated it at 0.1 Hz and 13.3 mA current [41]. Using acrylic and Teflon plates and nozzle diffuser design, they demonstrated a flowrate of 481.2 μL/min. Nam et al. [42] showed that IPMC membrane constrained on all sides didn't give desired deflections and the deflection could be increased further if this edge constraint was removed. So, they unconventionally used several unclamped actuators (instead of one) clamped at single edge and encapsulated between elastic films. This retained the Nafion water level and could sustain long operation times. Later, this group focussed on modifying the IPMC (Nafion and polyimide) for achieving higher deflection with low voltage [43]. Table 10.2 lists the recent work done in IPMC-based micropumps which highlights high displacements obtained at low voltages.

10.4 MATHEMATICAL MODEL FOR IPMC ACTUATOR

The mathematical model discussed below is based on theoretical works on IPMC membranes [45–47]. The ionic current in the polymer is calculated with the Nernst-Planck equation

$$\frac{\partial c}{\partial t} + \nabla . \left(-D\nabla C - z\mu FC\nabla \varphi - \mu CV_c \nabla P \right) = 0 \tag{10.1}$$

where C is cation concentration, μ is cation mobility, D is the diffusion coefficient of cation, F is Faraday's constant, z is the charge number of cation, V_c is the molar volume which quantifies the cation hydrophilicity, P is the solvent pressure and φ is the electric potential in the polymer region. The equation consists of the diffusive cation flux, migration in electric field and convective cation flux. The cation mobility

TABLE 10.2
IPMC Actuator-based Micropumps over Last 20 years

Reference	Type of work	Type of IPMC actuator	IPMC actuator shape	Dimensions	Micropump material and type	Voltage or current applied	Frequency applied	Maximum displacement	Flowrate (µL/min)
[44]	Fabrication	Platinum-coated Nafion 117 (4 membranes)	Circular, all sides constrained	Radius 8 mm	Acrylic, flap valve type	1.5 V	0.1 Hz	Not mentioned for multiple membranes	40 with 4 actuators
[36]	Fabrication	Nafion 112	Square, all sides constrained	8 mm*8 mm* 0.05 mm	PDMS, nozzle diffuser	2 V, 5 V	0.5 Hz	14 µm, 27 µm	9.97
[37]	Fabrication	Silver plated Nafion	Cantilever, single side constrained	15 mm*5 mm	NM*	1 V, 2 V, 3 V	1 Hz	2.1 mm, 4.4 mm, 6.78 mm	
[38]	Theoretical & simulation	Platinum-coated circle-shaped IPMC diaphragm with circular shaped electrode	Circular, all sides constrained	Radius 10 mm with electrode radius 8.5 mm	NM	2 V	0.1 Hz	0.966 mm	NM
[39]	Simulation	Platinum-coated Nafion	Circular, square, all sides constrained	Radius 10 mm, square area 314.159 mm²	NM	2 V	0.1 Hz	0.686 mm (circular), 0.760 mm (square)	NM

NM – not mentioned

* NM – not mentioned

is given by $\mu = D/(R*T)$. Here, R is the universal gas constant and T is the temperature. In case of actuation, the electric potential gradient term is significantly more prevalent than the solvent pressure flux, that is $zF\nabla\varphi \gg VC\nabla P$. The electric potential gradient term can be described using Poisson's equation as follows:

$$-\nabla^2\varphi = \frac{\rho}{\varepsilon} \qquad (10.2)$$

Where, ρ is defined as the space charge density in the Nafion part given by $\rho = (C - C_a)$. Here, C_a is the anion concentration and is related to the initial cation concentration, C_o, as:

$$C_a = C_0\left[1 - \left(\frac{\partial u_1}{\partial x} + \frac{\partial u_2}{\partial y} + \frac{\partial u_3}{\partial z}\right)\right] \qquad (10.3)$$

For most applications, it is reasonable to assume that $C_a = C_o$. The linear elastic material model is used to describe the IPMC deformation. The Navier's equation of displacement is used to solve the solid mechanics of the model. The system is said to be in equilibrium if $-\nabla.\sigma = F_v$ is satisfied. Here F_v is the body force per unit volume. This is due to the migration of cations towards the cathode and for a 2D model it is approximately related to cation concentration as:

$$F_v = \alpha*(C - C_a) \qquad (10.4)$$

For a 3D model, the body force is related to the cation concentration as:

$$F_v = \alpha*(C - C_a) + \beta*(C - C_a)^2 \qquad (10.5)$$

where, α and β are the constants for the linear term and the quadratic term respectively. For evaluating electrical current in electrodes, following equations are applied.

$$\nabla.J = Q_j \qquad (10.6)$$

$$J = \sigma*E + \frac{\partial D}{\partial t} + J_e \qquad (10.7)$$

$$E = -\nabla V \qquad (10.8)$$

where, J is the electric current, E is the electric field, σ is the electrical conductivity of gold, V is the voltage field vector, J_e is the externally generated current density and D is the electric displacement field vector. The electric field is obtained through the last equation. The electric potential V is defined by $V = V_0$ where V_0 is the applied voltage at the top and bottom electrodes respectively. These equations form the basis of finite element method-based simulation discussed in the next section.

10.5 EXAMPLE 10.1 – SIMULATION OF IPMC ACTUATOR CLAMPED ON ONE SIDE

The IPMC actuator follows transduction mechanism which is governed by multiphysics nature and requires multiphysics platform as employed in earlier simulation-based works [38,39,48]. The existing physics-based electromechanical transduction, or IPMC actuation models consider ionic current as the main component that causes the actuation. The model works well in the low voltage range for regular IPMCs. The underlying cause of IPMC transduction is ion migration and resulting spatial charge density in the vicinity of the electrodes. The IPMC actuator considered for simulation is shown in shown in Figure 10.2.

Four physics modules and four physics study interfaces were applied for designing the IPMC actuator. The methodology adopted is shown in Figure 10.3. The output of one physics module or interface determines the input to the next module or interface irrespective of the domains they are applied on. The output of one physics module or interface determines the input to the next module or interface irrespective of the domains they are applied on. The modules that are used in the study are the AC/DC module, the mathematics module, the structural mechanics module, and the chemical species transport module. The electric currents interface within the AC/DC module is used to simulate the current in a conductive and capacitive material under the influence of an electric field.

10.5.1 ELECTRIC CURRENTS INTERFACE

The electric currents interface within the AC/DC module is used to simulate the current in a conductive and capacitive material under the influence of an electric field.

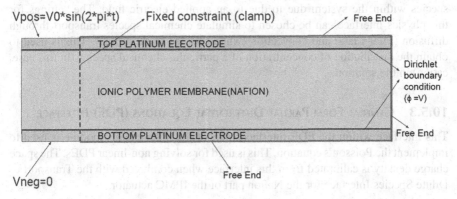

FIGURE 10.2 Two-dimensional view of IPMC actuator model considered for simulation with boundary conditions applied.

TABLE 10.3
Nafion Membrane Parameters Considered for Simulation

S/n	Parameter	Symbol	Value
1	IPMC membrane width	Width_IPMC	10 mm
2	IPMC thickness		0.127 mm
3	Electrode material		Gold
4	Electrode thickness		0.008 mm
5	Cation diffusion constant	D_cat	$0.7*10^{-11}[m^2/s]$
6	Initial upstream concentration of cation	C_cat	$1200[mol/m^3]$
7	Valency of cation	z_cat	1
8	Universal gas constant	R	8.314 [J/K/mol]
9	Dielectric permittivity	epsilon	2 [mF/m]
10	Faraday's constant	Farad	96485.3415 [C/mol]
11	Young's modulus of IPMC diaphragm	Young_IPMC	41.2 [MPa]
12	Poisson's ratio of IPMC	Poisson_IPMC	0.49
13	Applied voltage	Sine wave	5 V
14	Density of IPMC diaphragm	density_IPMC	$2000 [kg/m^3]$
15	Linear coefficient	Alpha	0.0001 [N/C]
16	Temperature	T	300 [K]
17	Time (for time-dependent studies)		8 s (step 0.1 s)

The electric potential in the desired domain is evaluated from the equations of current conservation. It solves for the electric potential in the electrode region of the model. In this case the electric potential is the input control signal which causes the actuation mechanism, i.e. deformation.

10.5.2 TRANSPORT OF DILUTED SPECIES INTERFACE

This interface deals with diffusion, convection as well as migration of the chemical species within the system due to due to an applied electric field. The settings for this physics interface can be chosen to simulate chemical species transport through diffusion (Fick's law) and convection (when coupled to fluid flow). This is used to obtain the distribution of concentration of a particular chemical species (in this case, cation) in the solution.

10.5.3 GENERAL FORM PARTIAL DIFFERENTIAL EQUATIONS (PDE) INTERFACE

This interface within the PDE interfaces part of the mathematics module is used to implement the Poisson's equation. This is used for solving non-linear PDEs. The space charge density is calibrated from this interface when combined with the Transport of Dilute Species Interface for the Nafion part of the IPMC actuator.

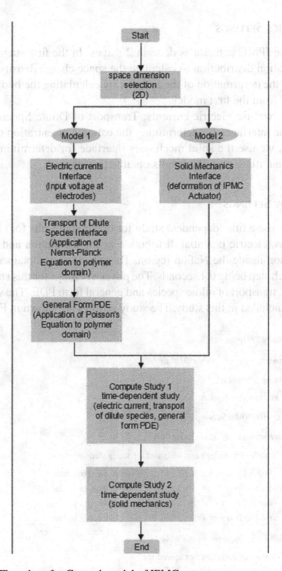

FIGURE 10.3 Flowchart for Comsol model of IPMC actuator.

10.5.4 Solid Mechanics Interface

The solid mechanics interface within the structural mechanics module solves for deformation based on stress analysis. For our modelling, we use the default linear elastic material to compute the displacement of the IPMC actuator in Y-direction due to the boundary load on the model, which is related to the space charge density, calculated in the previous interface. The linear elastic material uses a linear elastic equation for displacements based on user-defined linear elastic properties of the material like Young's modulus, Poisson's ratio and density of the IPMC diaphragm.

10.5.5 Physics Settings

Modelling of the IPMC actuator is done in 2 stages. In the first stage we model the cation concentration distribution to calculate the space charge density. In the second stage we model the deformation of the actuator by calculating the body load from the results obtained from the first model.

For model 1, we use electric currents, Transport of Dilute Species and General Form PDE as the interfaces for determining the cation concentration distribution.

For model 2, we use the solid mechanics interface for determining the displacement of the actuator based on the results obtained in model 1.

10.5.6 Study Settings

For model 1, we use a time-dependent study for obtaining results for electric potential on the electrodes, electric potential distribution inside the Nafion and cation concentration distribution inside the Nafion region. These results are obtained for the first 8 seconds with each step being 0.1 seconds. The physics settings for this study are only for electric currents, transport of dilute species and general form PDE. The solid mechanics interface is not included in this study. The study settings are shown in Figure 10.4.

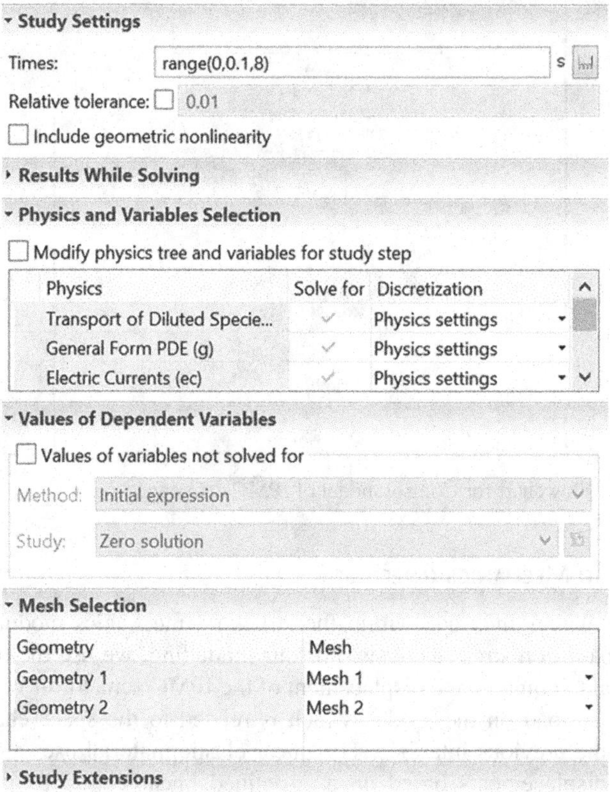

FIGURE 10.4 Study settings for model 1.

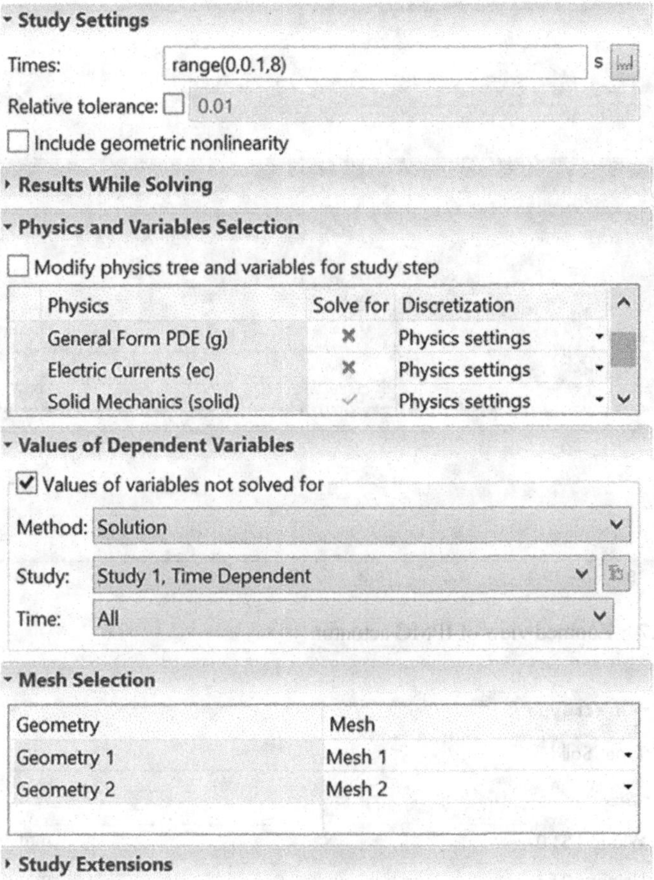

▾ **Study Settings**

Times: range(0,0.1,8)   s

Relative tolerance: ☐ 0.01

☐ Include geometric nonlinearity

▸ **Results While Solving**

▾ **Physics and Variables Selection**

☐ Modify physics tree and variables for study step

Physics	Solve for	Discretization	⌃
General Form PDE (g)	✖	Physics settings ▾	
Electric Currents (ec)	✖	Physics settings ▾	
Solid Mechanics (solid)	✓	Physics settings ▾	⌄

▾ **Values of Dependent Variables**

☑ Values of variables not solved for

Method: Solution   ⌄

Study: Study 1, Time Dependent   ⌄

Time: All   ⌄

▾ **Mesh Selection**

Geometry	Mesh	
Geometry 1	Mesh 1	▾
Geometry 2	Mesh 2	▾

▸ **Study Extensions**

FIGURE 10.5 Study settings for model 2.

FIGURE 10.6 Geometry of IPMC actuator.

For model 2, we also use a time-dependent study for obtaining results for displacement. This result is also obtained for the first 8 seconds with each step being 0.1 seconds. The physics settings for this study only include the solid mechanics interface. The study settings for model 2 are shown in Figure 10.5.

10.5.7 GEOMETRY

We use 2D space dimension for modelling the IPMC diaphragm. The geometry of IPMC actuator is shown in Figure 10.6 and Figure 10.7. The bottom rectangle and the top rectangle represent the platinum (Pt) electrodes of the IPMC actuator. The centre rectangle represents the Nafion polymer region of the diaphragm. The fourth

0.9
0.8
0.7
0.6
0.5
0.4
0.3
0.2
0.1
0
-0.1
-0.2
-0.3
9.2
9.4
9.6
9.8
10
10.2
10.4
10.6
10.8

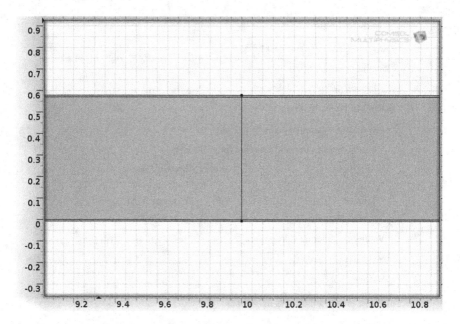

FIGURE 10.7 Zoomed view of IPMC actuator.

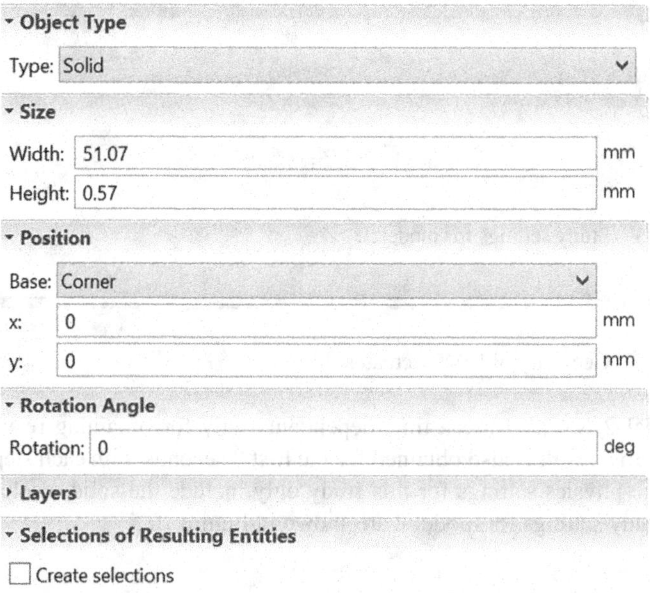

Object Type
Type: Solid
Size
Width: 51.07 mm
Height: 0.57 mm
Position
Base: Corner
x: 0 mm
y: 0 mm
Rotation Angle
Rotation: 0 deg
Layers
Selections of Resulting Entities
Create selections

FIGURE 10.8 Settings for r1 (Nafion region).

rectangle is used to clamp the actuator at a distance of 3.9 mm from the left extreme end. The dimensions of each component used for the geometry is represented in Figure 10.8 (for r1), Figure 10.9 (for r2), Figure 10.10 (for r3), Figure 10.11 (for r4), Figure 10.12 (for pt1) and Figure 10.13 (for pt2).

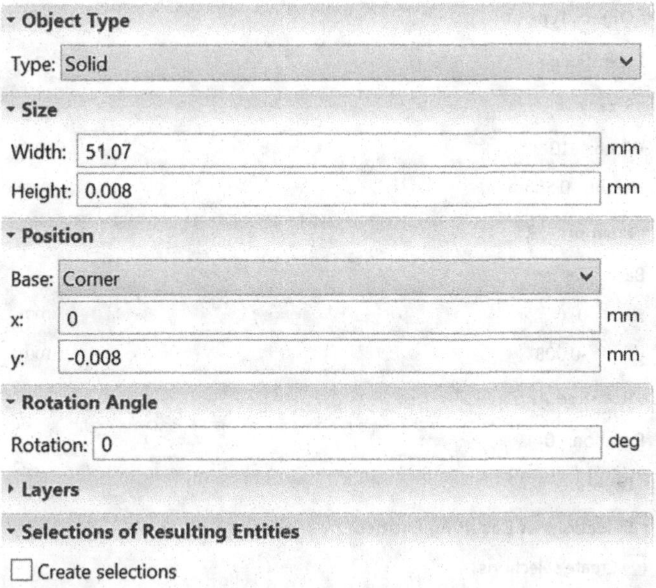

- **Object Type**

Type: Solid

- **Size**

Width: 51.07   mm

Height: 0.008   mm

- **Position**

Base: Corner

x: 0   mm

y: -0.008   mm

- **Rotation Angle**

Rotation: 0   deg

- **Layers**

- **Selections of Resulting Entities**

☐ Create selections

FIGURE 10.9 Settings for r2 (Bottom Pt electrode).

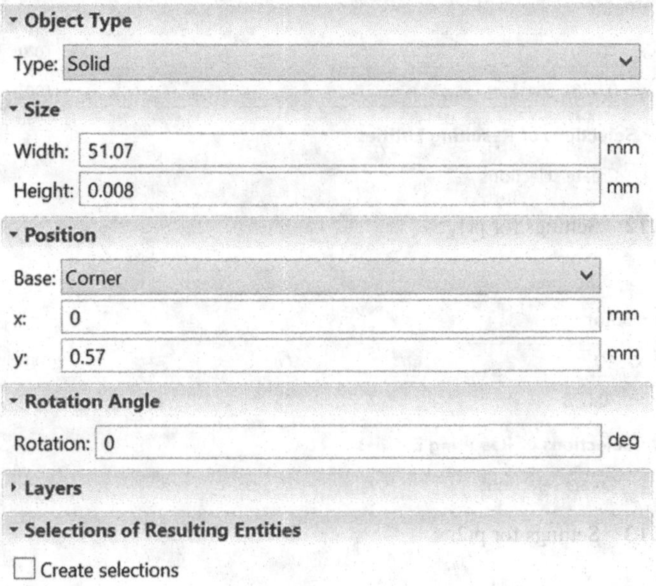

- **Object Type**

Type: Solid

- **Size**

Width: 51.07   mm

Height: 0.008   mm

- **Position**

Base: Corner

x: 0   mm

y: 0.57   mm

- **Rotation Angle**

Rotation: 0   deg

- **Layers**

- **Selections of Resulting Entities**

☐ Create selections

FIGURE 10.10 Settings for r3 (Top Pt electrode).

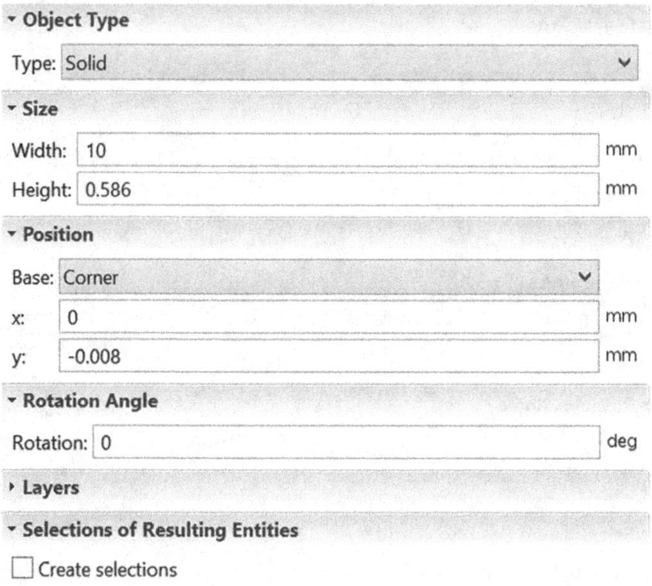

▾ Object Type

Type: Solid   ▾

▾ Size

Width: 10   mm

Height: 0.586   mm

▾ Position

Base: Corner   ▾

x: 0   mm

y: -0.008   mm

▾ Rotation Angle

Rotation: 0   deg

▸ Layers

▾ Selections of Resulting Entities

☐ Create selections

FIGURE 10.11   Settings for r4.

▾ Point

x: 10   mm

y: 0.578   mm

▾ Selections of Resulting Entities

☐ Create selections

FIGURE 10.12   Settings for pt1.

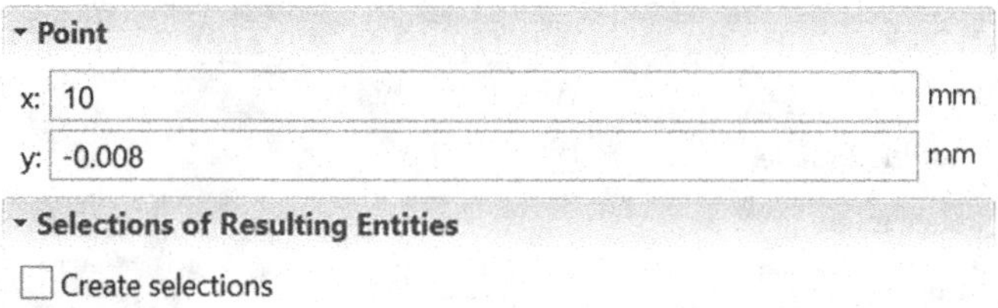

▾ Point

x: 10   mm

y: -0.008   mm

▾ Selections of Resulting Entities

☐ Create selections

FIGURE 10.13   Settings for pt2.

10.5.8   GLOBAL PARAMETERS AND VARIABLES

The parameters and the variables which are globally declared for the model are given in Table 10.4 and Table 10.5 respectively.

The variables declared for model 1 and model 2 are given in Table 10.6 and Table 10.7.

TABLE 10.4
Global Parameters for IPMC Diaphragm

Name	Expression	Description
D_cat	$0.7*10^{-11}$ [m²/s]	Diffusion constant of cation
C_cat	1200 [mol/(m³)]	Initial upstream concentration of cation
z_cat	1	Valency of cation
R	8.314 [J/K/mol]	Universal gas constant
epsilon	2 [mF/m]	Dielectric permittivity
Farad	96485.3415 [C/mol]	Faraday's constant
Young_IPMC	41.2 [MPa]	Young's modulus of IPMC diaphragm
Poisson_IPMC	0.49	Poisson's ratio of IPMC diaphragm
density_IPMC	2000 [kg/(m³)]	Density of IPMC diaphragm
Alpha	0.0001 [N/C]	Linear coefficient
T	300 [K]	Temperature
V0	5 [V]	Maximum electrode potential for Vpos
width_IPMC	10 [mm]	Thickness of IPMC diaphragm

TABLE 10.5
Global Variables for IPMC Diaphragm

Name	Expression	Unit	Description
u_cat	D_cat/(R*T)	s.mol/kg	Mobility of cation
Vpos	V0*sin(2*pi[rad/s]*t)[V]	V	Electrode potential
Vneg	0 [V]	V	Ground

TABLE 10.6
Variables for Model 1

Name	Expression	Unit	Description
rho	Faraday*(c-C_cat)	C/(m³)	Charge density
resist	1/(1.1125 [ohm^-1])*(2 [cm])	Ohm-m	1/(electrical conductivity of electrodes)

TABLE 10.7
Variables for Model 2

NAME	EXPRESSION	UNIT	DESCRIPTION
F_z	Alpha*mod1.idmap1(mod1.rho)	N/(m³)	Body force

10.5.9 Electric Current Module

The electric currents module is used to determine the flow of electric current in the platinum electrodes. Thus, domains 1, 3, 4 and 6 are selected and the following equations are applied in order to obtain the solutions for this model:

$$\nabla.J = Q_j$$
$$j = \sigma * E + \frac{\partial D}{\partial t} + J_e$$
$$E = -\nabla V$$

Where, J is the electric current, E is the electric field, σ is the electrical conductivity of platinum, V is the voltage field vector, J_e is the externally generated current density, D is the electric displacement field vector, The electric field is obtained through the last equation.

The electric potential V is defined by

$$V = V_0$$

Where V_0 is the applied voltage at the top and bottom electrodes respectively.

These are the required equations which are solved for current conservation of platinum electrodes. These set of equations solve for the electric current flow in the electrode domains. The interfacial boundaries between the electrodes and polymer membrane are insulated. The initial value of electric potential V is assigned as 0[V]. Electric potential 1 of V_0 =Vpos is assigned to the top electrode (boundary 5 and boundary 7). Vpos has been declared in the global variables. Electric potential 2 of V_0 =Vneg is applied to the bottom electrode (boundary1 and boundary 2). Vneg has also been defined in the global variables. The electrical conductivity and the relative permittivity values are user defined and are set as 1/resist and 1 respectively. Resist variable is declared in the variables for model 1. The out of the plane thickness is set as width_IPMC, taken from the global parameters. The settings are given in Figure 10.14, Figure 10.15, Figure 10.16 and Figure 10.17.

10.5.10 Transport of Diluted Species Module

The transport of dilute species interface solves for the migration of ions in the polymer domain. The cation concentration distribution throughout the Nafion region is determined from this interface. No flux condition is applied to the interfacial boundaries between the polymer membrane and electrodes. The diffusion and migration region is set to domain 2 and domain 5, i.e the Nafion domain only.

The Nernst-Planck equation is used to calculate the ionic current in the polymer and is given by:

$$\frac{\partial C}{\partial t} + \nabla \cdot (\underbrace{-D\nabla C}_{\substack{\text{diffusive} \\ \text{cation flux}}} - \underbrace{z\mu FC\nabla\phi}_{\substack{\text{migration} \\ \text{in elec. field}}} - \underbrace{\mu C V_C \nabla P}_{\substack{\text{convective} \\ \text{cation flux}}}) = 0$$

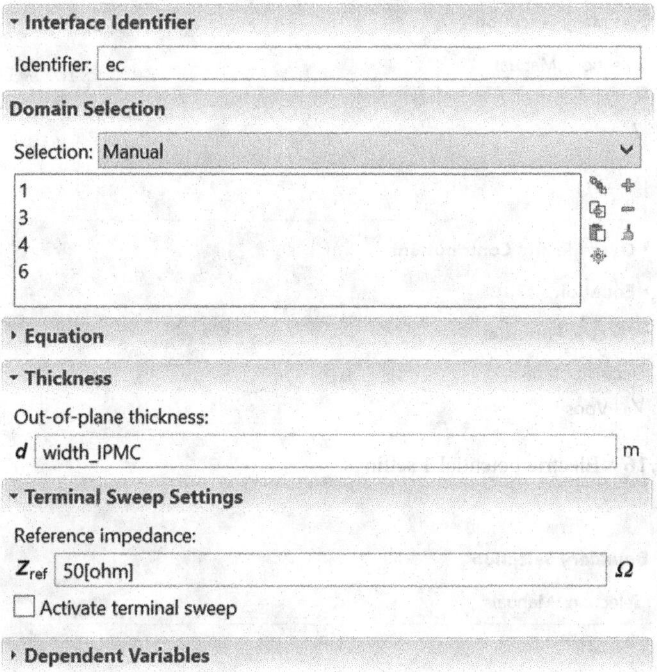

Interface Identifier

Identifier: ec

Domain Selection

Selection: Manual

1
3
4
6

▸ **Equation**

▾ **Thickness**

Out-of-plane thickness:

d width_IPMC   m

▾ **Terminal Sweep Settings**

Reference impedance:

Z_{ref} 50[ohm] Ω

☐ Activate terminal sweep

▸ **Dependent Variables**

FIGURE 10.14 Electric currents settings.

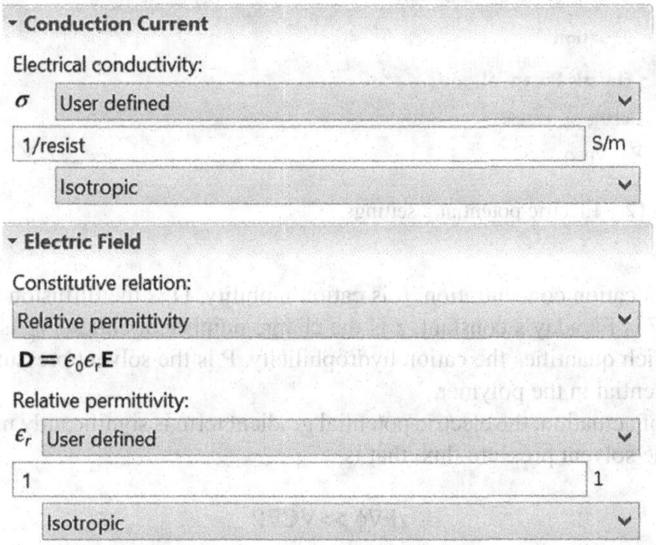

▾ **Conduction Current**

Electrical conductivity:

σ User defined   ⌄

1/resist   S/m

Isotropic   ⌄

▾ **Electric Field**

Constitutive relation:

Relative permittivity   ⌄

$D = \epsilon_0 \epsilon_r E$

Relative permittivity:

ϵ_r User defined   ⌄

1   1

Isotropic   ⌄

FIGURE 10.15 Current conservation settings.

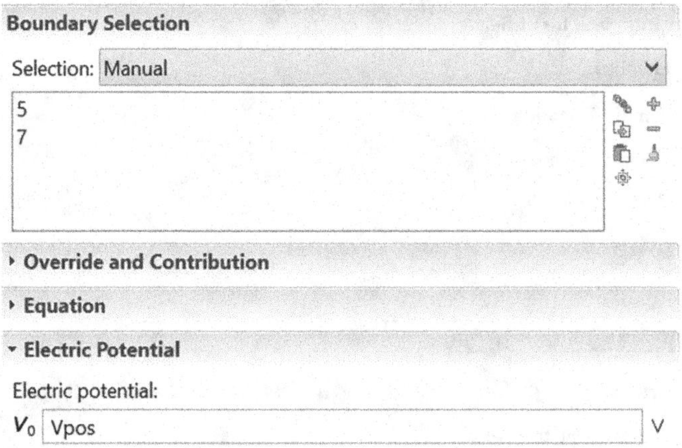

FIGURE 10.16   Electric potential 1 settings.

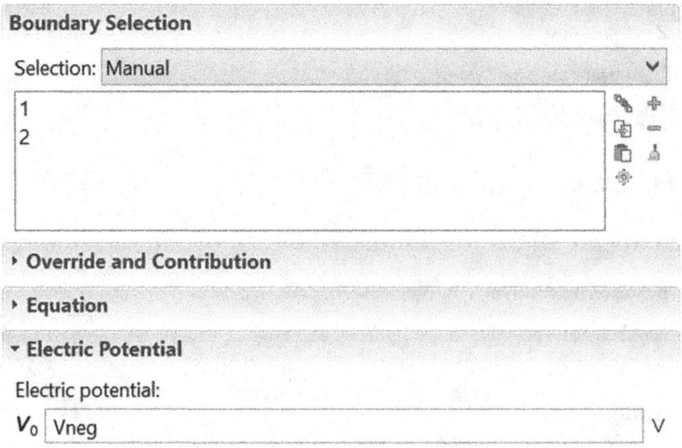

FIGURE 10.17   Electric potential 2 settings.

Where: C is cation concentration, μ is cation mobility, D is the diffusion coefficient of cation, F is Faraday's constant, z is the charge number of cation, V_C is the molar volume which quantifies the cation hydrophilicity, P is the solvent pressure, ϕ is the electric potential in the polymer.

In case of actuation, the electric potential gradient term is significantly more prevalent than the solvent pressure flux, that is

$$zF\nabla\phi \gg VC\nabla P$$

So, we neglect the convective cation flux term in the NP equation. For a time-dependent study, the equations used in Comsol modelling are as follows:

$$\frac{\partial c_i}{\partial t} + \nabla.\left(-D_i\nabla c_i - z_i u_{m,i} F c_i \nabla V\right) = R_i$$

$$N_i = -_i\nabla c_i - z_i u_{m,i} F c_i \nabla V$$

This is comparable with the approximated NP equation discussed above for time-dependent study.

Here, R_i is the reaction term particularized for each ion and in our case, it is 0, comparing it with the NP equation. D_i is the diffusivity of cation which has already been declared in the global parameters as D_cat, c_i is the cation concentration, z_i is the valence of cation whose value is z_cat, $u_{m,i}$ is the cationic mobility given by my_cat in the global variables, F is Faraday's constant, declared as Farad in global parameters, $\Phi = V$ represents the potential field vector in the Nafion part.

N_i denotes the mass flux vector of the cationic species.

The diffusion coefficient, mobility and charge number are user defined as set as, D_cat, my_cat and z_cat respectively which have been previously declared in the parameters and variables. The initial cation concentration is set as C_cat. The term V in the equations represents the electric potential inside the polymer region, which is set as u, which is defined by the Poisson's equation in the General Form PDE interface. Figure 10.18, Figure 10.19 and Figure 10.20 represent the settings for this interface.

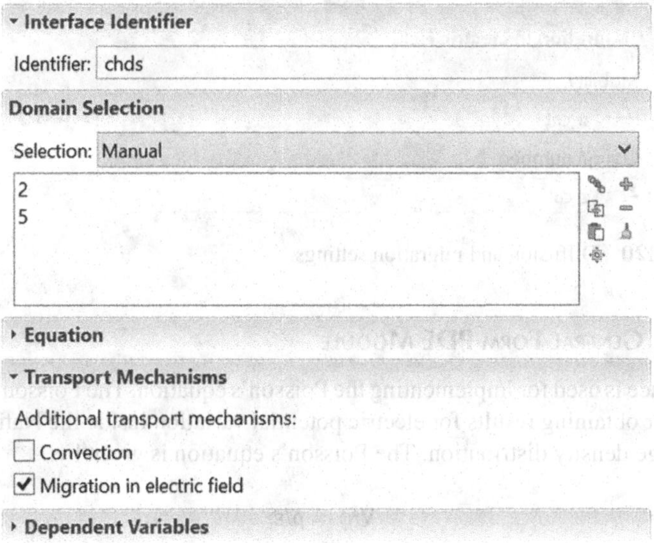

FIGURE 10.18 Transport of diluted species.

⏷ Initial Values

Concentration:

c | C_cat | mol/m³

FIGURE 10.19   Initial values settings.

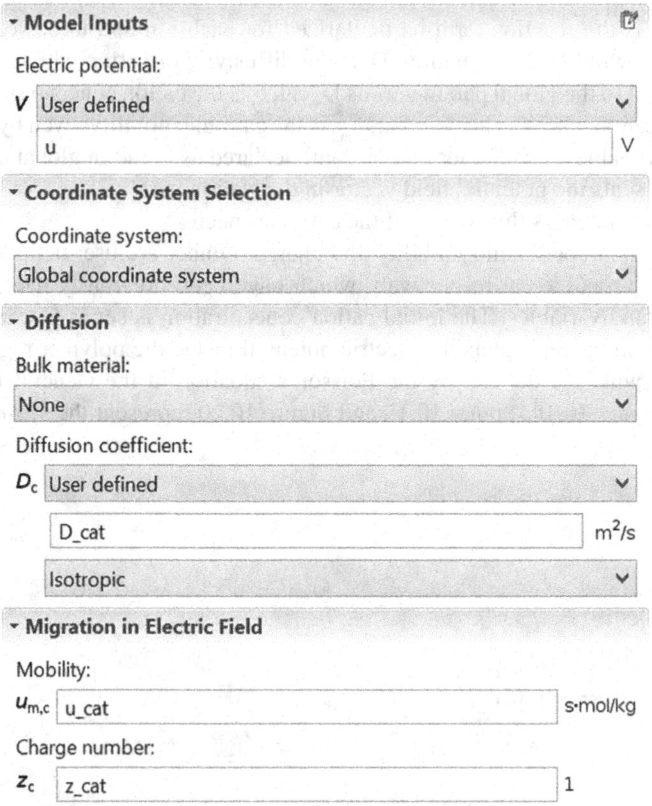

⏷ Model Inputs

Electric potential:

V | User defined ⌄

| u | V

⏷ Coordinate System Selection

Coordinate system:

| Global coordinate system | ⌄

⏷ Diffusion

Bulk material:

| None | ⌄

Diffusion coefficient:

D_c | User defined | ⌄

| D_cat | m²/s

| Isotropic | ⌄

⏷ Migration in Electric Field

Mobility:

$u_{m,c}$ | u_cat | s·mol/kg

Charge number:

z_c | z_cat | 1

FIGURE 10.20   Diffusion and migration settings.

10.5.11   GENERAL FORM PDE MODULE

This interface is used for implementing the Poisson's equation. The Poisson's equation is useful for obtaining results for electric potential variation inside the Nafion and the space charge density distribution. The Poisson's equation is given by:

$$-\nabla^2 \phi = \rho/\varepsilon$$

Here, ε is epsilon which defined in the global parameters.

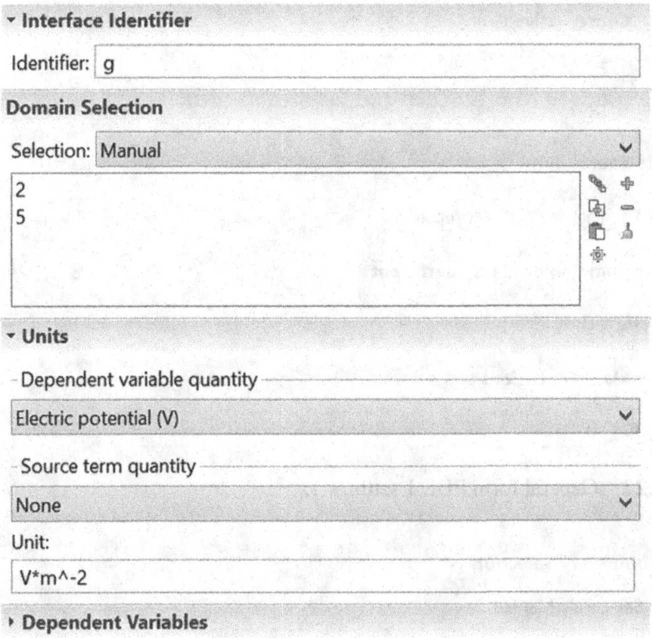

Interface Identifier

Identifier: g

Domain Selection

Selection: Manual

2
5

Units

Dependent variable quantity

Electric potential (V)

Source term quantity

None

Unit:

V*m^-2

▸ **Dependent Variables**

FIGURE 10.21 General form PDE settings.

The space charge density, ρ is given by the following expression:

$$\rho = Faraday*(c-C\_cat)$$

Here, c is the cation concentration calibrated from the previous interface, C_cat is the initial cation concentration and Faraday refers to the Faraday's constant.

This interface is applied for domains 2 and 5. The dependent variable quantity is set as electric potential and the unit is changed to $V*m^{-2}$. The source term is given by Faraday*(c-C_cat)/epsilon. The damping coefficient is set to 0. Dirichlet boundary conditions are applied to the boundaries 4,6,11,13 and the prescribed value of u is set to V. The settings for General Form PDE are given in Figure 10.21, Figure 10.22 and Figure 10.23.

10.5.11 Solid Mechanics Module

This interface is used to calculate the tip displacement of the IPMC actuator by using the results of space charge density calibrated using the previous three models. The body load on the actuator is responsible for causing deformation which is almost linearly dependent on the space charge density created due to cation mobility towards the cathode. A linear elastic material model has been applied. The equations used in this module are as follows:

$$-\nabla.\sigma = Fv$$

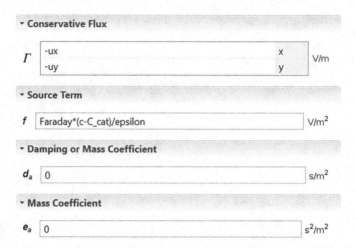

FIGURE 10.22   General form PDE 1 settings 1.

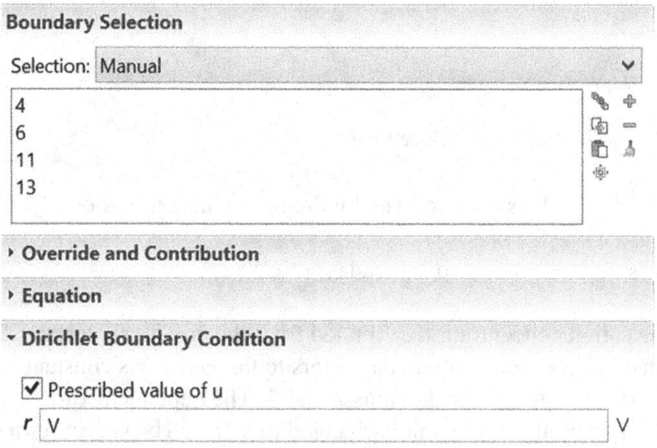

FIGURE 10.23   Dirichlet boundary condition settings.

Here, F_v stands for body load and has been declared in the variables as F_z. For a 2D rectangular model, body load is given by

$$F\_z = Alpha*rho$$

Where Alpha and rho (space charge density) are declared in the global parameters and variables respectively.

The thickness is taken to be width_IPMC. The Young's modulus, Poisson's ratio and density of the selected domains is user defined as Young_IPMC, Poisson_IPMC, density_IPMC. These terms are declared in the global parameters.

This interface is added to the domains 2,4,5 and 6. The domains, 8,10 and 12 are clamped and thus, a fixed constraint is applied to it.

The boundary load 1 is applied to boundary 13 and the value of load in Y-direction is expressed as -(F_z*width_IPMC). The boundary load 2 is applied to boundary 11 and the value of load in Y-direction is expressed as (F_z*width_IPMC).

Figure 10.24, Figure 10.25, Figure 10.26 and Figure 10.27 show the settings for solid mechanics.

10.5.12 Simulation Results

10.5.12.1 Result for Electric Potential at Electrodes

The 2D plot for electric potential in the electrodes is shown in the figure for time= 0.3 s. The top electrode has a uniform potential of 0.4757 V while the bottom electrode

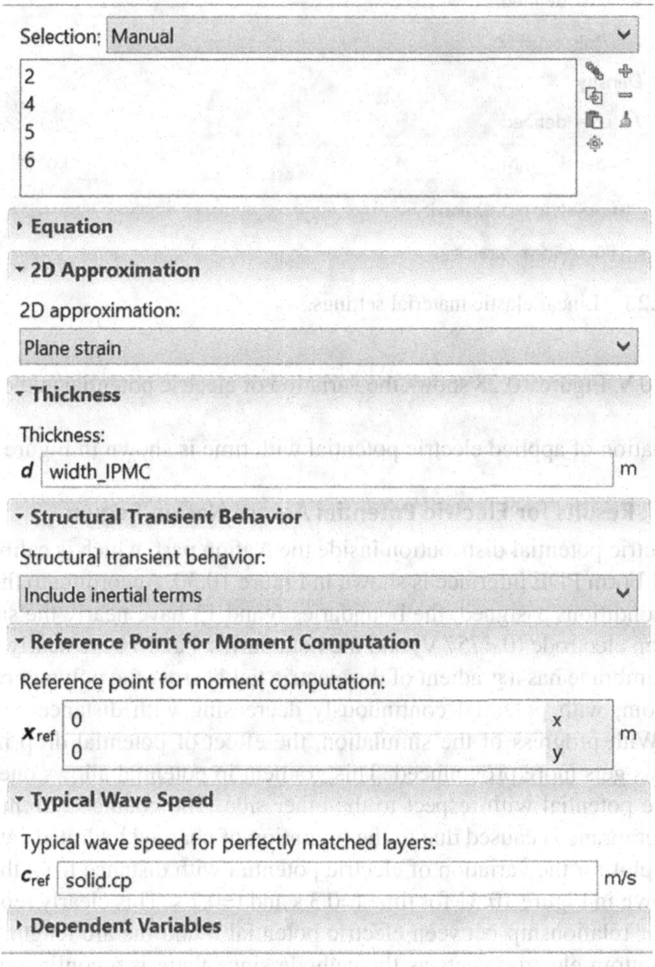

FIGURE 10.24 Solid mechanics settings.

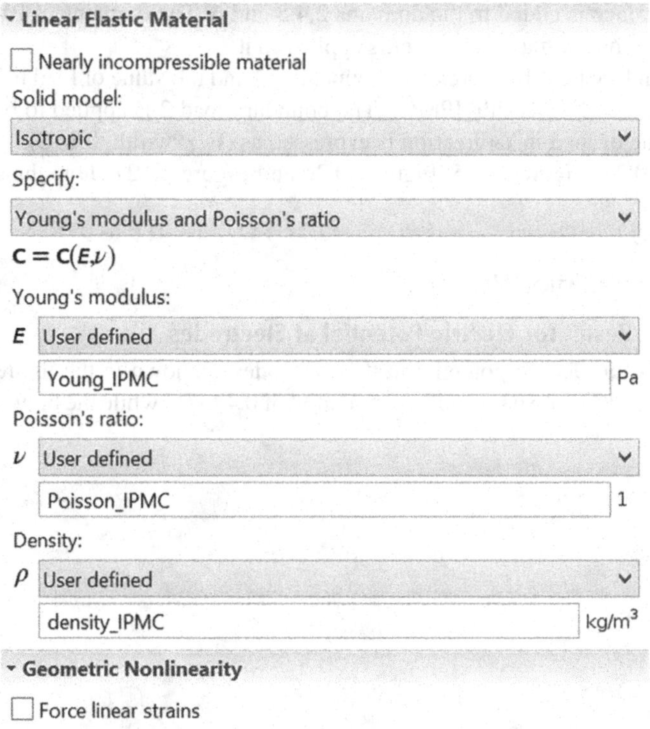

▾ Linear Elastic Material

☐ Nearly incompressible material

Solid model:

Isotropic	∨

Specify:

Young's modulus and Poisson's ratio	∨

$C = C(E,\nu)$

Young's modulus:

E	User defined	∨

	Young_IPMC	Pa

Poisson's ratio:

ν	User defined	∨

	Poisson_IPMC	1

Density:

ρ	User defined	∨

	density_IPMC	kg/m^3

▾ Geometric Nonlinearity

☐ Force linear strains

FIGURE 10.25   Linear elastic material settings.

is nearly at 0 V. Figure 10.28 shows the variation of electric potential at the electrodes with time.

The variation of applied electric potential with time is shown in Figure 10.29.

10.5.12.2   Results for Electric Potential Across Nafion Region

The 2D electric potential distribution inside the Nafion part, which is calibrated from the General Form PDE interface is shown in Figure 10.30. According to the Dirichlet boundary conditions assigned, the boundaries 6 and 13 have nearly the same potential as the top electrode (0.4757 V) and the boundaries 4 and 11 are nearly at 0 V. The polymer membrane has a gradient of the electric field across the polymer region from top to bottom, with potential continuously decreasing with distance from the top electrode. With progress of the simulation, the effect of potential drop in the electrode regions gets more pronounced. This gradient in potential allows one side to be at a positive potential with respect to the other side. The counter current inside the polymer membrane is caused due to the migration of charged hydrated cations.

The 1D plot for the variation of electric potential with distance from the top electrode is shown in Figure 10.31 for time t=0.3 s and t=0.7 s. This clearly represents the nearly linear relationship between electric potential u and the arc length. For time= 0.3 s, the bottom electrode acts as the cathode since there is a continuous decrease

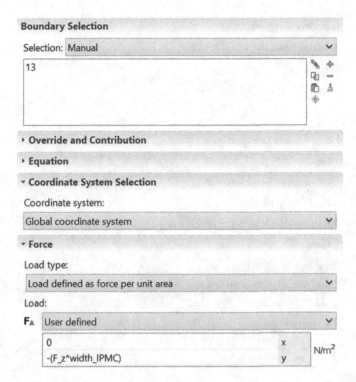

FIGURE 10.26 Boundary load 1 settings.

in electric potential from the top to bottom electrode. For t=0.7 s, the top electrode acts as the cathode since the bottom electrode is at nearly 0 V and the top electrode is nearly at -0.4754 V.

10.5.12.3 Results for Distribution of Cation Concentration

The 2D cation concentration distribution inside the Nafion part is shown in the Figure for t=0.3 s. The transport of dilute species takes place only in the Nafion region. The electrodes are exempted from this because there is no migration of ions in the electrode regions. Initially, there was a uniform cation concentration of 1200 mol/m³ throughout the Nafion membrane. When the results of electric potential were fed to this interface, a drastic change in the model is visualized. Since the cations are positively charged, they tend to migrate towards the more negative electrode. For time= 0.3 s, the bottom electrode is nearly at 0 V while the top electrode is at 0.4757 V. So, the hydrated cations migrate towards the bottom electrode and most of the cations accumulate near the cathode end (in this case, bottom electrode is the cathode). The concentration near the anode end starts decreasing and thus a gradient in concentration is noticed when there is a gradient in the electric field in the Nafion region. This variation in cation concentration ranges from 517.46 mol/m³ near the anode end to about 2321.4 mol/m³ near the cathode end. The cation concentration across the Nafion region is visualized in Figure 10.32 and Figure 10.33.

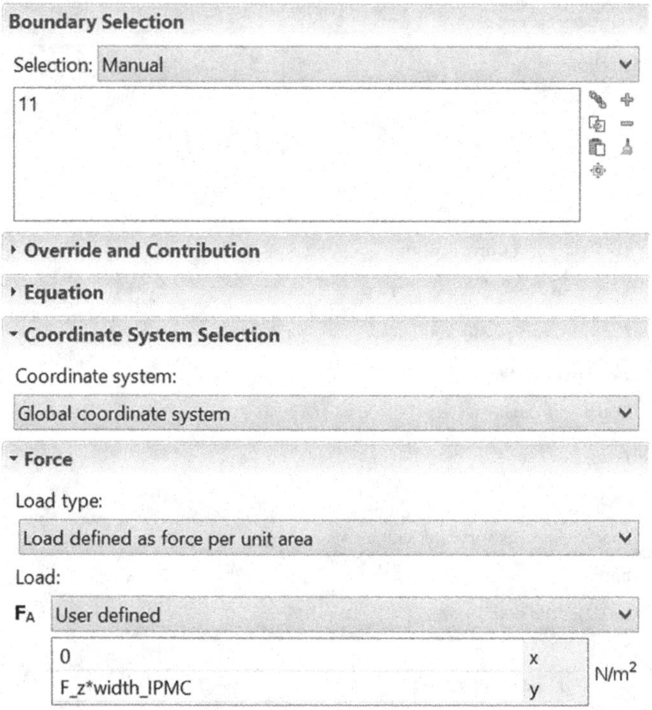

Boundary Selection

Selection: Manual

11

▸ **Override and Contribution**

▸ **Equation**

▾ **Coordinate System Selection**

Coordinate system:

Global coordinate system

▾ **Force**

Load type:

Load defined as force per unit area

Load:

F_A User defined

0	x	
F_z*width_IPMC	y	N/m^2

FIGURE 10.27 Boundary load 2 settings.

From the 1D plot for the variation of cation concentration across the Nafion region in Figure 10.34, we observe that there is a sudden decrease in the cation concentration near the anode and a sudden increase in cation concentration near the cathode. These correspond to the anode boundary layer and the cathode boundary layer respectively. Here 0mm arc length corresponds to anode and the 0.57 mm corresponds to cathode for t=0.3 s. When time=0.7 s, 0 mm arch length corresponds to cathode interface and 0.57 mm corresponds to anode interface. Due to this reason, now the concentration at 0.57 mm has decreased from 2321.4 mol/m^3 to 1986.3 mol/m^3 while the cation concentration at 0 mm has increased from 517.46 mol/m^3 to 653.56 mol/m^3. This agrees with the double layer theory behind cation migration.

The cation concentration at the bottom polymer-electrode interface varied with time when study was computed for 8 seconds with a step of 0.1 seconds. As time changes constantly, the magnitude and polarity of the applied voltage varies sinusoidally, resulting in sinusoidal change in the cation concentration accordingly as seen in the Figure 10.35.

A parametric sweep was applied to V0sin(2πt) where t=0.3 s and V0 was made to vary from 0 V to 5 V with each step being 0.5 V. At t=0.3 s, the bottom electrode acts as the cathode. The plot of variation of cation concentration at cathode vs electric potential (Vpos) is shown in Figure 10.36. It is seen that, when a strong electric field

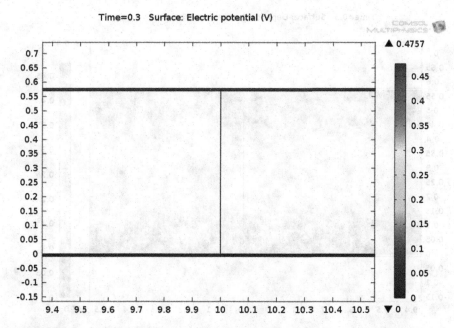

FIGURE 10.28   Electric potential distribution in Pt electrodes.

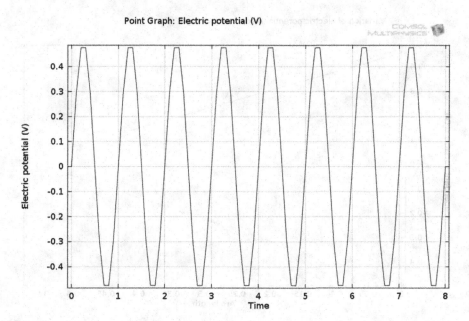

FIGURE 10.29   Variation of electric potential (Vpos).

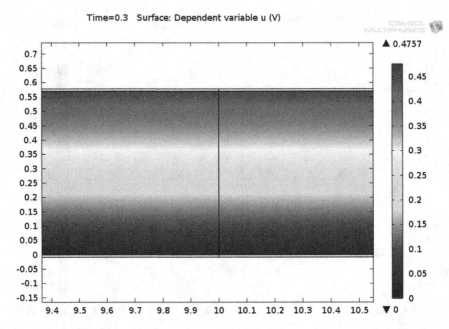

FIGURE 10.30   Potential gradient across the Nafion.

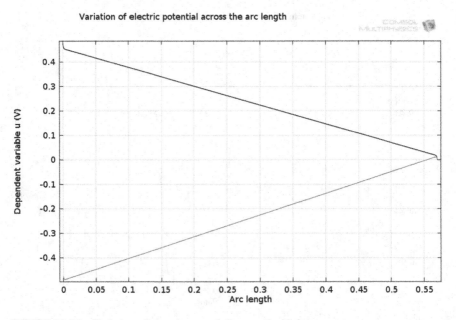

FIGURE 10.31   Variation of electric potential across the Nafion.

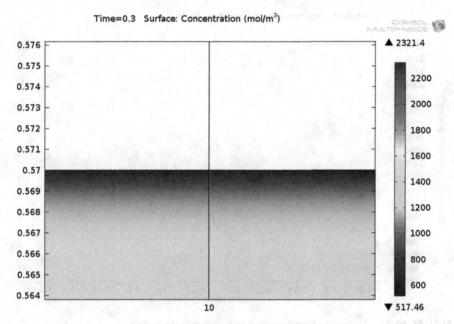

FIGURE 10.32 Concentration gradient near anode boundary.

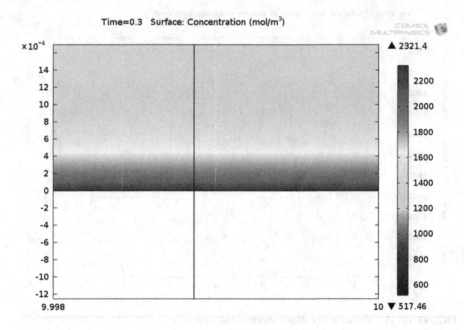

FIGURE 10.33 Concentration gradient near cathode boundary.

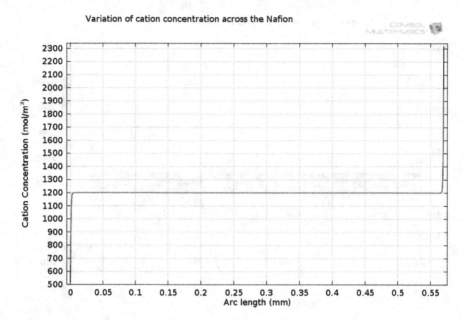

FIGURE 10.34   Variation of cation concentration across the Nafion.

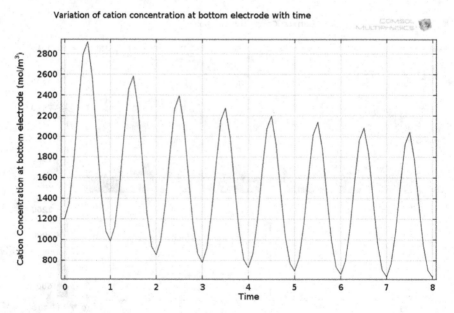

FIGURE 10.35   Variation of cation concentration with time.

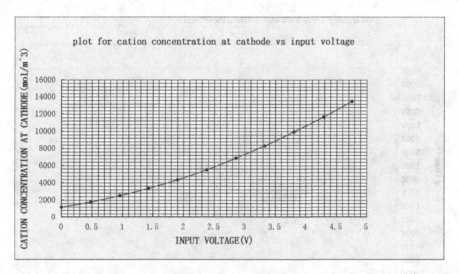

FIGURE 10.36   Variation of cation concentration with input voltage.

is applied, the cation concentration near the cathode increases exponentially, resulting in stronger deflection.

10.5.12.4   Results for Deformation of IPMC Actuator

The body load varies linearly with cation concentration. This is seen in Figure 10.37 at time=0.3 s.

The variation of body load across the Nafion region can be seen from the plot for time=0.3 s in Figure 10.38. Close to the anode boundary, Body load is negative due to reduced cation concentration. Near the cathode boundary it increases linearly with distance due to increase in cation concentration. In the remaining part, it is almost 0 due to nearly constant cation concentration.

Figure 10.39 shows the Von Mises stress for the actuator at time=0.3 s. The maximum value is found to be 1.9541 MPa, i.e. near the clamped region. Deformation takes places only in Y-direction.

The plot for variation of maximum displacement in Y-direction along the length of the IPMC Diaphragm (Figure 10.40) shows that till 10 mm, it is clamped and hence there is no displacement. After that, displacement in Y-direction increases exponentially with distance from the clamped region. At time=0.3 s, the displacement is nearly 61.14327 mm at the tip. The plot (Figure 10.41) shows the variation of displacement along Y-direction with time. On comparing the displacement versus time plot with that of applied electric potential versus time plot, it can be observed that when potential increases displacement increases. Moreover, displacement also depends on the polarity of the electric potential.

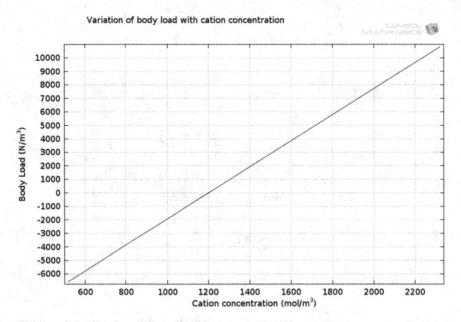

FIGURE 10.37 Variation of body load with cation concentration.

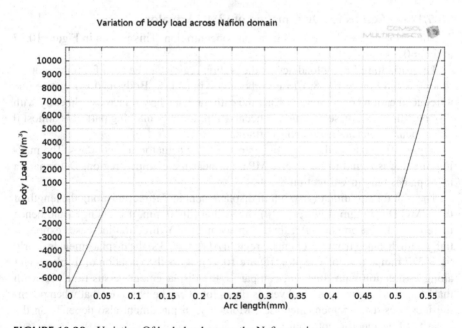

FIGURE 10.38 Variation Of body load across the Nafion region.

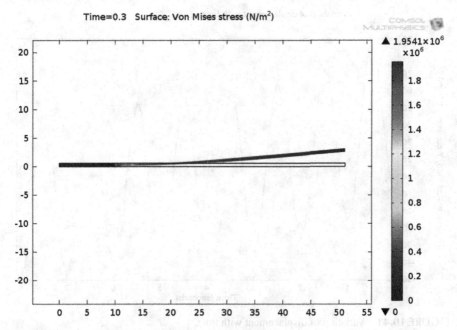

FIGURE 10.39 Von Mises stress.

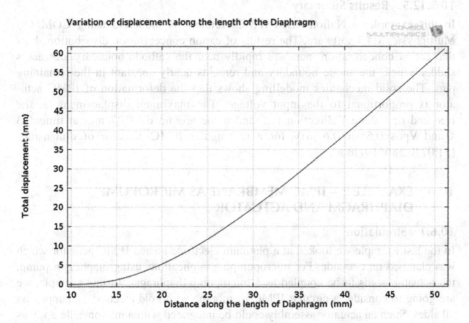

FIGURE 10.40 Variation of displacement along the diaphragm length.

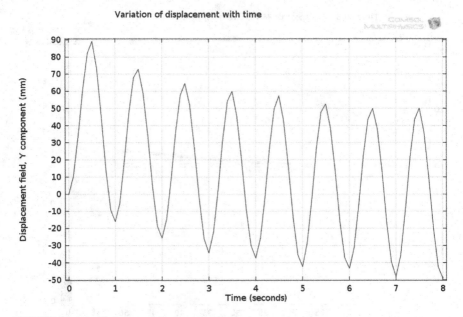

FIGURE 10.41 Variation of displacement with time.

10.5.12.5 Results Summary

In this example, a Nafion-based IPMC Actuator is modelled using COMSOL Multiphysics v.4.3 software. The results of cation concentration distribution show that cation concentration increases rapidly near the cathode boundary, decreases suddenly near the anode boundary and remains nearly constant in the remaining part. The solid mechanics modelling shows that the deformation of IPMC actuator is proportional to the input voltage. The maximum displacement (at the free end tip) along Y-direction is found to be around 6.14327 mm at time=0.3 s and Vpos=0.5*sin(0.6*pi)V for a rectangular IPMC actuator of dimensions $(51.07*0.586*10)$ mm^3.

10.6 EXAMPLE 2 – IPMC MEMBRANE AS MICROPUMP DIAPHRAGM AND ACTUATOR

10.6.1 Simulation

In the last example we looked at a platinum electrode coated IPMC actuator which was clamped on one side. For micropumping applications using diaphragm pump, the actuator needs to be coupled as a micropump diaphragm. In this example, we are going to consider a circular IPMC membrane with gold electrodes clamped on all sides. Such an actuator assembly could be integrated with a microneedle array as discussed in previous chapters to have a micropump and microneedle integrated drug delivery system. The proposed schematic of such a system is shown in Figure 10.42. Here, we shall confine our discussion to 2-dimensional representation. The simulation

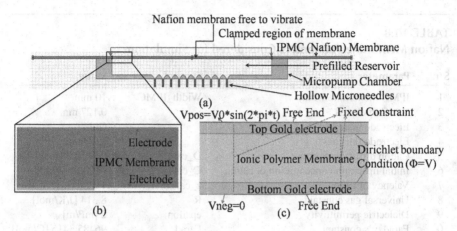

FIGURE 10.42 Two-dimensional view of IPMC actuator model considered for simulation with boundary conditions applied.

approach remains the same and the modules are addressed as discussed in the last example. This section shall briefly discuss the approach adopted and simulation results obtained.

The electric potential in the desired domain is evaluated from the equations of current conservation. It solves for the electric potential in the electrode region of the model. The settings for this physics interface can be chosen so as to simulate chemical species transport through diffusion (Fick's law) and convection (when coupled to fluid flow). This is used to obtain the distribution of concentration of a particular chemical species (in this case, cation) in the solution. General form partial differential equations (PDE) interface within the PDE interfaces part of the mathematics module is used to implement the Poisson's equation.

This is used for solving non-linear PDEs. The space charge density is calibrated from this interface when combined with the Transport of Dilute Species Interface for the Nafion part of the IPMC actuator. The solid mechanics interface within the structural mechanics module solves for deformation based on stress analysis. For modelling, we have used the default linear elastic material to compute the displacement of the IPMC actuator in Y-direction due to the boundary load on the model, which is related to the space charge density, calculated in the previous interface. The linear elastic material uses a linear elastic equation for displacements based on user-defined linear elastic properties of the material like Young's modulus, Poisson's ratio and density of the IPMC diaphragm. Table 10.8 gives the parameter values considered for simulation.

The polymer membrane has a gradient of the electric field across the polymer region from top to bottom, with potential continuously decreasing with distance from the top electrode. This gradient in potential allows one side to be at a positive potential with respect to the other side. The counter current inside the polymer membrane is caused due to the migration of charged hydrated cations. The plot for the variation of electric potential with distance from the top electrode to the bottom electrode at

TABLE 10.8
Nafion Membrane Parameters Considered for Simulation

S/n	Parameter	Symbol	Value
1	IPMC membrane width	Width_IPMC	10 mm
2	IPMC thickness		0.127 mm
3	Electrode material		Gold
4	Electrode thickness		0.008 mm
5	Cation diffusion constant	D_cat	$0.7*10^{-11}$ [m^2/s]
6	Initial upstream concentration of cation	C_cat	1200 [mol/ m^3]
7	Valency of cation	z_cat	1
8	Universal gas constant	R	8.314 [J/K/mol]
9	Dielectric permittivity	epsilon	2 [mF/m]
10	Faraday's constant	Farad	96485.3415 [C/mol]
11	Young's modulus of IPMC diaphragm	Young_IPMC	41.2 [Mpa]
12	Poisson's ratio of IPMC	Poisson_IPMC	0.49
13	Applied voltage	Sine wave	3 V
14	Density of IPMC diaphragm	density_IPMC	2000 [kg/m^3]
15	Linear coefficient	Alpha	0.0001 [N/C]
16	Temperature	T	300 [K]
17	Time (for time-dependent studies)		8 s (step 0.1s)

different time is shown in Figure 10.43. It represents the nearly linear relationship between electric potential and the arc length (cutline covering distance between top and bottom electrode).

When an external field is applied to the IPMC membrane, the hydrated cations migrate towards the cathode and most of the cations accumulate near the cathode end. The concentration near the anode end starts decreasing and thus a gradient in concentration is noticed when there is a gradient in the electric field in the membrane region. This can be observed in Figure 10.44a where the cation surface concentration ($1933X10^3$ mol/m^3) at the bottom electrode drops to $672X10^3$ mol/m^3. From the plot for the variation of cation concentration across the IPMC membrane region at different time intervals for applied 0.5 volts in Figure 10.44b, we observe again that there is a sudden decrease in the cation concentration near the anode and a sudden increase in cation concentration near the cathode. These correspond to the anode boundary layer and the cathode boundary layer respectively. This agrees with the double layer theory behind cation migration. Figure 10.45 shows the Von Mises stress for the IPMC actuator clamped at both ends at time=8 s. The stress reaches a maximum value of around 0.05 Mpa near the constrained edges and at the centre. The membrane will not be permanently damaged till the stress value does not cross the yield strength of Nafion (43 Mpa). Deformation takes places only in Y-direction. Figure 10.46 shows the response of the IPMC membrane (centre point of the circular membrane precisely) to a sinusoidal 3 V input with time. Figure 10.47 gives the plot of the applied signal and corresponding displacement of the IPMC membrane over

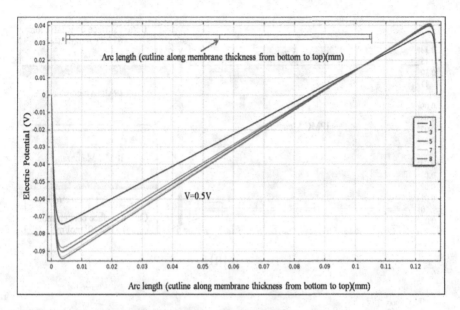

FIGURE 10.43   Variation of electric potential across IPMC membrane.

its length. The deformation closely follows the cation distribution affected by applied sinusoidal voltage. The results are summarized in Table 10.9.

10.6.2   RESULTS SUMMARY

In this example, IPMC membrane-based actuator was designed and simulated. Cation concentration, stress and deflection etc. were estimated using numerical analysis and simulation. More analysis is required to strengthen the simulation model for IPMC by bringing in the damping introduced by viscosity of practical fluids.

10.7   CONCLUSION

In this chapter, IPMC actuators were explored as an micropump actuator alternative. IPMC membrane-based actuator was designed and simulated. Cation concentration, stress and deflection etc. were estimated using numerical analysis and simulation. More analysis is required to strengthen the simulation model for IPMC by bringing in the damping introduced by viscosity of practical fluids.

Hence, we could summarize that IPMC is a candid material that can be used for the fabrication of the actuator unit of the micropump. IPMC actuators overcome the above-mentioned disadvantages due to other actuators as observed below:

- The electromechanically actuated IPMC has the ability to create larger deformations (over 1% of bending strain).
- It requires a very low input voltage (1–2 V) in order to exhibit large displacements.

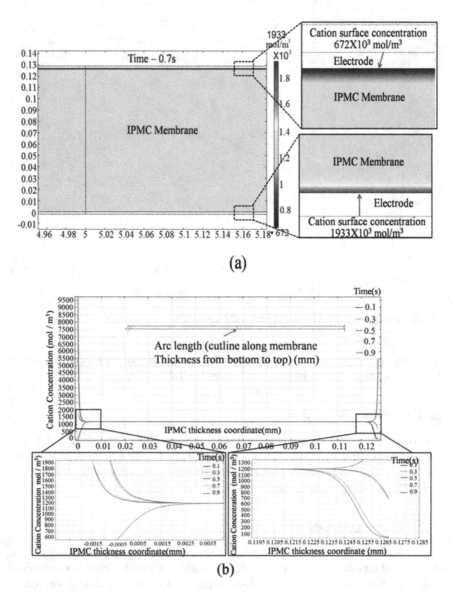

FIGURE 10.44 Variation of Von Mises stress across the IPMC membrane length at time= 0.35 seconds. The stress reaches a maximum value of around 5 Mpa near the constrained edges and at the centre. Deflection of centre of IPMC membrane is 260 μm at 3 V at 0.5 seconds. Deflection of the centre point of the IPMC membrane with applied sinusoidal voltage (0.5 V) at different time intervals.

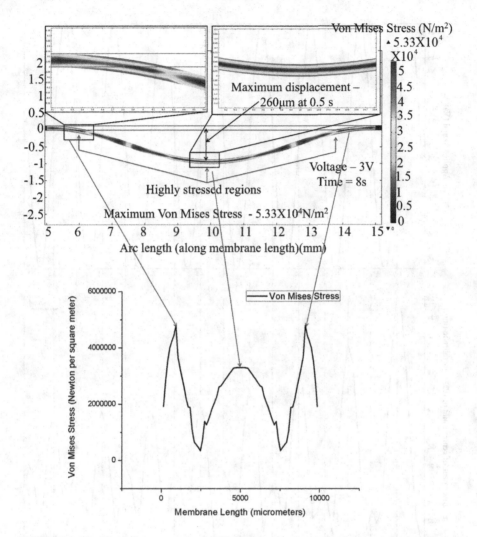

FIGURE 10.45 Von Mises stress for the IPMC actuator clamped at both ends at time=8 s.

- It shows fairly a quick response (>120 Hz).
- It is versatile and can be fabricated in various shapes and sizes.
- It is bio-compatible and soft.
- It doesn't generate excessive heat or electromagnetic waves and has low power consumption.
- It can operate not only in liquid but also in air.

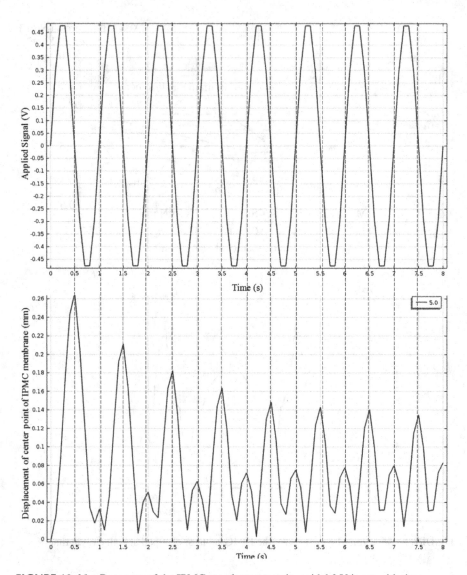

FIGURE 10.46   Response of the IPMC membrane to a sinusoidal 3 V input with time.

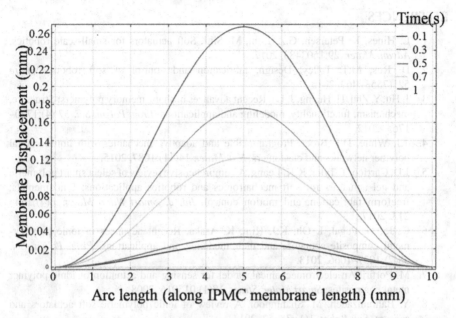

FIGURE 10.47 Plot of the applied signal and corresponding displacement of the IPMC membrane over its length.

TABLE 10.9
Results for Example 10.2 IPMC Membrane Simulation

S/n	Parameter	Value
	At 3 V, 0.5 Hz, 0.7 seconds	
1	Cation concentration near top electrode	672×10^3 mol/m³
2	Cation concentration near bottom electrode	1933×10^3 mol/m³
3	Maximum Von Mises stress	0.053 Mpa or 5.33×10^4 N/m²
4	Highly stressed regions of IPMC membrane	Boundary where membrane is constrained and centre of the membrane
5	Displacement	0.12 mm
6	Maximum displacement	0.26 mm at 0.5 s
7	Response time	0.25 s
8	Displacement direction	Positive

REFERENCES

1. L. Hines, K. Petersen, G.Z. Lum, M. Sitti, Soft actuators for small-scale robotics, *Advan. Mater.* 29:1603483, 2017.
2. D. Rus, M.T. Tolley, Design, fabrication and control of soft robots, *Nature* 521(7553):467, 2015.
3. J. Hu, Y. Zhu, H. Huang, J. Lu, Recent advances in shape memory polymers: Structure, mechanism, functionality, modeling and applications, *Prog. Polym. Sci.* 37 (4):1720–1763, 2012.
4. T.J. White, D.J. Broer, Programmable and adaptive mechanics with liquid crystal polymer networks and elastomers, *Nat. Mater.* 14(11):1087, 2015.
5. J.D. Carrico, T. Tyler, K.K. Leang, A comprehensive review of select smart polymeric and gel actuators for soft mechatronics and robotics applications: fundamentals, freeform fabrication, and motion control, *Int. J. Smart Nano Mater.* 8(4):144–213, 2017.
6. C. Jo, D. Pugal, I. Oh, K.J. Kim, K. Asaka, Recent advances in ionic polymer-metal composite actuators and their modeling and applications, *Prog. Polym. Sci.* 38(7):1037–1066, 2013.
7. M. Porfiri, An electromechanical model for sensing and actuation of ionic polymer metal composites, *Smart Mater. Struct.* 18(1):015016, 2008.
8. Y. Bahramzadeh, M. Shahinpoor, A review of ionic polymeric soft actuators and sensors, *Soft Robot.* 1(1):38–52, 2014.
9. R. Tiwari, E. Garcia, The state of understanding of ionic polymer metal composite architecture: A review, *Smart Mater. Struct.* 20(8):083001, 2011.
10. C.H. Wirguin, Recent advances in perfluorinated ionomer membranes: structure, properties and applications, *J. Membr. Sci.* 120:1–33, 1996.
11. K.A. Mauritz, R.B. Moore, State of understanding of Nafion, *Chem. Rev.* 104:4535–86, 2004.
12. T.D. Gierke, G.E. Munn, F.C. Wilson, Morphology of perfluorosulfonated membrane products wide-angle and small-angle x-ray studies, *J. Polym. Sci.: Polym. Phys. Ed.* 19:1687–1704, 1981.
13. T.D. Gierke, G.E. Munn, F.C. Wilson, Perfluorinated ionomer membrane, chapter 10 – The Morphology in nafion perfluorinated membrane products, as determined by wide- and small-angle x-ray studies. *Am. Chem. Soc.* 19:1687–1704, 1982.
14. K.J. Kim, M. Shahinpoor, Ionic polymer-metal composites: II. manufacturing techniques, *Smart Mater. Struct.* 12(1):65–79, 2003.
15. V. Palmre, D. Pugal, K. J. Kim, K.K. Leang, K. Asaka, A. Aabloo, Nanothorn electrodes for ionic polymer-metal composite artificial muscles, *Sci. Rep.* 4(6176):1–10, 2014.
16. P. de Gennes, K. Okumura, M. Shahinpoor, K.J. Kim, Mechanoelectric effects in ionic gels, *Europhys. Letters* 50:513–518, 2000.
17. Y. Cha, M. Porfiri, Mechanics and electrochemistry of ionic polymer metal composites, *J. Mech. Phys. Solids.* 71:156–178, 2014.
18. I.S. Park, S.M. Kim, D. Pugal, L. Huang, S.W. Tam-Chang, K.J. Kim, Visualization of the cation migration in ionic polymer–Metal composite under an electric field, *Appl. Phys. Lett.* 96(4):043301, 2010.
19. N. Lakshminarayanaiah, *Transport phenomena in membranes*, New York, NY: Academic Press, 1969.
20. J. Lee, J. Nam, H. Choi, H.M. Kim, J. Jeon, H.K. Kim, Water uptake and migration effect on IPMC (ion-exchange polymer metal composite) actuator. *Proc. SPIE.* 4329:84–92, 2001.

21. M. Shahinpoor, K.J. Kim. Ionic polymer–metal composites: III. Modeling and simulation as biomimetic sensors, actuators, transducers, and artificial muscles. *Smart Mater. Struct.* 13:1362–1388, 2004.
22. S.N. Nasser, C. Thomas. Ionomeric polymer – metal composites electroactive polymer (EAP) actuators as artificial muscles. *Reality, Potential and Challenges*, Bellingham, WA: SPIE Optical Engineering Press chapter 6, 171–230, 2004.
23. T. Satoshi, S. Yamagami, T. Takamori, K. Oguro, Modeling of Nafion-Pt composite actuators (ICPF) by ionic motion. *SPIE's 7th Annual International Symposium on Smart Structures and Materials*, 92–102, Newport Beach, CA, 2000.
24. K. Mallavarapu, M.N. Kenneth, D.J. Leo. Feedback control of the bending response of ionic polymer-metal composite actuators. *SPIE's 8th Annual International Symposium on Smart Structures and Materials*, 301–310, Newport Beach, CA, 2001.
25. K.M. Newbury, D.J. Leo. Linear electromechanical model of ionic polymer transducers-part I: model development. *Journal of Intelligent Material Systems and Structures.* 14(6):333–342, 2003.
26. K. Farinholt, D.J. Leo. Modeling of electromechanical charge sensing in ionic polymer transducers. *Mech. Mater.* 36(5):421–433, 2004.
27. N. Nasser, Sia, J. Yu. Li. Electromechanical response of ionic polymermetal composites. *J. Appl. Phys.* 87(7):3321–3331, 2000.
28. N. Nasser, Siavouche, Y. Wu. Tailoring actuation of ionic polymer metal composites through cation combination. *Smart Struct. Mater. Int. Soc. Opt. Photonics*, 245–253, 2003. DOI: 10.1117/12.484439
29. N. Nasser, Sia, S. Zamani, Y. Tor. Effect of solvents on the chemical and physical properties of ionic polymer-metal composites. *J. Appl. Phys.* 99(10):104902, 2006.
30. S. Lee, H.C. Park, K.J. Kim. Equivalent modeling for ionic polymer–metal composite actuators based on beam theories. *Smart Mater. Struct.* 14(6):1363, 2005.
31. P.J.C. Branco, J.A. Dente. Derivation of a continuum model and its electric equivalent-circuit representation for ionic polymer–metal composite (IPMC) electromechanics. *Smart Mater. Struct.* 15(2):378–392, 2006.
32. D.J. Leo, K. Farinholt, T. Wallmersperger, Computational models of ionic transport and electromechanical transduction in ionomeric polymer transducers. *Smart Struct. Mater.* 5759:170–181, 2005.
33. T. Wallmersperger, D.J. Leo, C.S. Kothera. Transport modeling in ionomeric polymer transducers and its relationship to electromechanical coupling. *J. Appl. Phys.* 101(2):024912, 2007.
34. T. Wallmersperger, B.J. Akle, D.J. Leo, B. Kröplin, Electrochemical response in ionic polymer transducers: An experimental and theoretical study. *Compos. Sci. Technol.* 68(5):1173–1180, 2008.
35. X. He, D.J. Leo, T. Wallmersperger. Modeling of ion transport in high strain ionomers by Monte Carlo simulation compared to continuum model. *Proc. ASME Int. Mech. Eng. Congr. Expo.* (Paper number IMECE2006-13928) 71:119–126, 2006.
36. J.J. Pak et al., Fabrication of ionic-polymer-metal-composite (IPMC) micropump using a commercial Nafion, *Proc. SPIE 5385, Smart Struct. Mater. - Electroactive Polym. Actuat. Dev. (EAPAD)* – San Diego, CA, United Kingdom, 2004.
37. C.K. Chung, P.K. Fung, Y.Z. Hong, M.S. Ju, C.C.K. Lin, T.C. Wu, *A Novel Fabrication of Ionic Polymer-Metal Composites (IPMC)Actuator with Silver Nano-Powders, The 13th International Conference on Solid-State Sensors, Actuators and Microsystems*. Digest of Technical Papers. TRANSDUCERS '05, Seoul, South Korea, 2005.
38. S. Lee and K.J Kim, Design of IPMC actuator-driven valveless micropump and its flow rate estimation at low Reynolds numbers, *Smart Mater. Struct.* 15(4):1103–1109, 2006.

39. S. Lee et al., *Smart Ionic Polymer-Metal Composites: Design and their Applications in Progress in Smart Materials and Structures*, USA, 67–114, 2007.
40. T.T. Nguyen, N.S. Goo, V.K. Nguyen, Y.Yoo, S. Park, Design, fabrication, and experimental characterization of a flap valve IPMC micropump with a flexibly supported diaphragm, *Sensors Actuat. A: Phys.* 41(2):640–648, 2008.
41. J. Santos et al., Ionic polymer–metal composite material as a diaphragm for micropump devices, *Sensors Actuat. A: Phys.* 161(1–2): 225–233, June 2010.
42. D.N.C. Nam, K.K. Ahn, Design of an IPMC diaphragm for micropump application, *Sensors Actuat. A: Phys.* 187:174–182, 2012.
43. J. Nam, T. Hwang, K.J. Kim, D.C Lee, A new high-performance ionic polymer–metal composite based on Nafion/polyimide blends, *Smart Mater. Struct.* 26(035015):11, 2017.
44. S. Guo, K. Asaka, Polymer-based new type of micropump for bio-medical application. *Proc. ICRA'03, IEEE Int. Conf. Robot. Autom.* 2:1830–1835, 2003.
45. M.W. Ashraf, S. Tayabba, N. Afzulpurkar, Micro electromechanical systems (MEMS) based microfluidic devices for biomedical applications, *Int. J. Mol. Sci.* 12:3648–3704, 2011.
46. T. Wallmersperger, D.J. Leo, C.S. Kothera, Transport modeling in ionomeric polymer transducers and its relationship to electromechanical coupling, *J. Appl. Phys.* 101(2):024912, 2007.
47. P. Nardinocchi, P. Matteo, L. Placidi, Thermodynamically based multiphysic modeling of ionic polymer metal composites, *J. Intell. Mater. Syst. Struct.* 22(16):1887–1897, 2011.
48. S. Ranjbarzadeh, *Modeling, Simulation and Applications of Ionic Polymer Metal Composites*, Ph.D. Thesis, Mechanical Engineering, COPPE, of the Federal University of Rio de Janeiro, 2017.

11 Microneedle and IPMC Micropump Integrated Drug Delivery System

11.1 INTRODUCTION

In the last few chapters, we learnt about microneedles and different actuation techniques for micropumps. Microneedles are micron sized pathways to skin while micropump-based drug delivery aims to achieve the therapeutic effect for a variety of drugs. We apply the knowledge gained from these chapters for the assembly of these components (microneedle, reservoir and micropump) and try to build a microneedle array integrated micropump system. This chapter discusses the integration approach with examples. Initially IPMC actuator-based micropump (particularly commercially available Nafion membrane) is discussed. Then we shall a step further in trying to define a probable fabrication approach for a drug delivery device on assembly of these components. As emphasized earlier, the fabrication path may vary depending upon the application, approach adopted and facilities available. The advantages of such a device shall lie in the precise control of the drug flowrate (where they are human compliant either at fast or slow rates) and extremely small size [1,2]. Micropump-based fluid delivery is either being used as a standalone device for drug delivery or maybe used in the lab-on-chip-based devices for diagnostics applications [3,4]. If we consider insulin delivery, flowrates of around 10–40 µL/min are required for bolus insulin delivery (i.e. insulin specifically taken at mealtime to keep blood glucose level under control) [5]. A micropump-based device targeting drug delivery should have low voltage operation, biocompatibility, portability, a wide range of flowrate and be light weight [2]. The device is explained with the help of three case studies/examples. They are:

(i) the first example shall examine the fabrication of the IPMC membrane and its characterization as a valveless nozzle diffuser structure-based micropump,
(ii) subsequently an insulin delivery device with microneedles and micropump is fabricated and characterized for tunable insulin delivery, and
(iii) the third case study presents an approach in IPMC membrane geometry modification to achieve high flowrate.

DOI: 10.1201/9781003202264-11

11.2 APPLICATION – INSULIN DELIVERY

An example of drug delivery device development for insulin delivery is considered in this chapter. The alarming number associated with diabetes has triggered the research and development for therapeutic products which blend into a patient's lifestyle. Diabetes is one of the most common lifelong disorders in children as well as adults. For the past several decades, there hasn't been a paradigm change in the options available for insulin delivery. Insulin is either taken several times a day through tablets, syringes, pens, inhalers or administered as a continuous subcutaneous infusion through an insulin pump. The hypodermic syringes are painful, traumatic and associated with frequent hypo-and hyperglycemia episodes due to unphysiological insulin delivery [6,7]. Their administration also requires training or trained medical staff. Insulin taken orally is destroyed in the gastrointestinal tract and is not effective. Similarly, insulin inhalers have also not been able to deliver insulin effectively to the body. Insulin pumps provide precise and regulated insulin delivery but are very expensive, and not acceptable to many patients, especially children and adolescents as they are continuously attached to the body. This has led diabetic patients to look for solutions which are small, painless, precise and self-administrable. Hence there is the requirement for a solution that can deliver insulin as precisely as pump infusion sets and at the same time provide painless a skin attachment/interphase like hollow microneedles. Another challenge lies in the transportation of the large insulin molecule (having molecular mass>5000 Da) across an outer layer of skin which cannot happen only by diffusion. For this, a micropump is required to push insulin into skin [3,8,9]. The next important parameter is to determine the kind of flowrates required for insulin delivery. The dose of insulin depends on several factors like age, weight, diet, physical activity etc. It can range from as low as 5 units per day in an infant to as high as 80 units or even higher in an overweight adolescent. We can narrow down the insulin dosage requirement based upon the flowrates which are offered by insulin pumps existing in market. The flow rate for basal insulin (longer acting insulin taken between meals) delivery if we are using U-100 insulin should be from 5–20 µL/min. It should also be able to deliver insulin boluses (shorter acting insulin taken after meals) at a rate varying from 20–100 µL/min [10,11].

11.3 INSULIN DELIVERY DEVICE

Based upon the discussions in the preceding chapters and the requirements discussed in the last section, an insulin delivery device (IDD) is proposed in this chapter having the following components:

  i.   an insulin reservoir,
  ii. a micropump actuated by Ionic Polymer metal composite (IPMC) membrane,
  iii. a hollow polymer microneedle array.

Feasibility studies need to be carried out on the proposed device and its components for different geometries of commercial Nafion membrane with the objective of determining its suitability for the IDD. The hollow SU-8 microneedle array can be fabricated by the direct laser writing technique as discussed in preceding chapters [12]. The microneedles may be fabricated on silicon wafer in which flow channels could be etched. The micropump can be positioned on top of the reservoir. The micropump actuator assembly can be made up of Nafion membrane sandwiched between two copper electrode rings connected to external electrical supply. On being electrically actuated, this Nafion membrane should deflect and press the liquid in the reservoir. With the proposed design, on each alternate cycle of Nafion membrane movement, insulin is pumped in from the reservoir and then pushed out through the microneedles. The micropump actuation assembly is the core component of this device.

11.5   EXAMPLE 11.1 – NAFION ACTUATOR-BASED MICROPUMP

11.5.1   BACKGROUND

After the fabrication of the first MEMS-based micropump in the 1990s, a lot of micropump techniques have been investigated with important parameters being actuation force, deformation, response time and reliability [13–18]. Out of the different micropump structures, one of the common structures of nozzle diffuser geometry is considered in this work to achieve the net flow from inlet to outlet [19]. Out of these, the popular piezoelectric actuators have the disadvantage of being heavy in weight, expensive in manufacturing and less compliant to be safely operated near humans [20,21]. This leads to exploration of smart membranes for pump which is beginning to become enabling technology for a micropump. The trend is towards the electrically driven ionic polymer metal composites (IPMC) membranes. Ionic polymer metal composites are generally ion exchange membranes which are sandwiched between noble metals such as gold or platinum as electrodes [22]. They have advantages of low actuation voltages, operability in aqueous environments, light weight and large range of deflections [23,24]. The electrode coating on IPMCs requires to be uniform, being strongly adhered to membrane and provide low electrical resistance [25]. The example discussed here is based on work by Mishra et al. [12].

11.5.2   FABRICATION

We shall look into micropump fabrication with the following parameters. What one will notice is that most of these parameters have been evolved from discussions in the previous chapters especially when dealing with micropump structure. We shall consider here a valveless nozzle diffuser design-based diaphragm pump. The parameters are shown in Table 11.1. A feasibility study is performed on the IPMC (Nafion 115) membrane to decide upon its suitability as micropump membrane. For this it is required to meet specifications of large displacements at low voltages.

11.5.3   FEASIBILITY STUDIES ON NAFION MEMBRANE

The IPMC membrane under consideration is Nafion. Nafion (perfluorovinyl ether groups terminated with sulfonate groups onto a tetrafluoroethylene backbone) is one of the readily commercially available IPMCs [26–28]. It is available in both liquid as

TABLE 11.1
Micropump Parameters for Fabrication

S/n	Parameter	Value
1	IPMC material	Nafion 115
2	IPMC membrane width	10 mm
3	IPMC membrane thickness	0.127 mm
4	Gold electrode thickness	0.008 mm
5	Micropump chamber material	PMMA
6	PMMA sheet thickness	1 mm
7	Polymer sheet thickness	100 μm
8	Micropump chamber depth in PMMA	100 μm
9	Rubber gaskets thickness	1.5 mm
10	Diffuser length	1700 μm
11	Diffuser inlet width	300 μm
12	Diffuser outlet width	750 μm
13	Divergence angle	10°
14	Distance between diffuser and outlet / inlet	8300 μm

well as sheet form (when cast into thin membranes). Depending upon the thickness of the membrane, Nafion is available as Nafion 112, 115, 117 etc. [29]. We shall consider Nafion 115 membrane for fabricating actuator assembly as it is the thinnest version available. Another approach could be to make Nafion membrane from solution as well. It has been observed that fabricated Nafion membrane with Li doping provides much better out-of-plane deflection compared to commercial Nafion-based membrane. Also synthesized Li doped Nafion membrane gives much better deflection in comparison to other membranes because of the ionic conductivity of the membrane is increased with increasing concentration of carrier cation (Li+) [30].

The proposed micropump structure is shown in Figure 11.1a. Commercially available Nafion 115 membrane (DuPont), US) is coated with gold (around 0.8 micron thickness by physical vapour evaporation). A thin interlayer of chromium (20 nm) was deposited to improve adhesion of gold on the Nafion membrane. Gold was heated by electrical resistive heating and condensed on the Nafion membrane. Here it is noteworthy to mention that out of different variants of Nafion membrane (like Nafion 112, 115 and 117) available from DuPont, Nafion 115 is found to be a better match for our application. In work by Guilly et al. [31] where the effect of Nafion membrane thickness on deformation was studied, it was found that unlike Nafion 115 (127 μm) or 117(183 μm), Nafion 112 (50 μm thick) membrane was unable to sustain the deformation induced by input signal due to its extreme flexibility and showed quick relaxation from the deformed position. Also, its thinness does not allow it to generate enough force at par with Nafion 115 and 117 (around 1.27/g) [31]. A higher thrust by the micropump membrane is required to pump liquid for such applications. Additionally, thicker Nafion membranes have higher operating lives [32]. Out of Nafion 115 and 117, Nafion 115 was chosen because of lesser thickness.

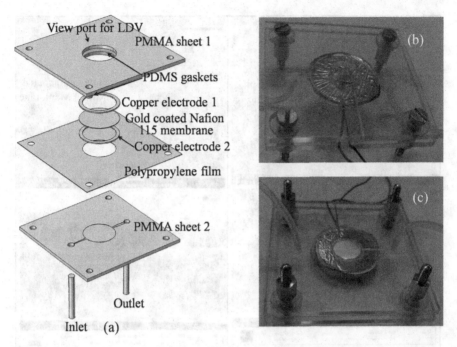

FIGURE 11.1 (a) Proposed schematic of the micropump with IPMC actuator assembly, (b) Top view of the device and (c) Bottom view of the fabricated device.

It's important to keep the Nafion membrane is proper orientation while carrying out the process of physical vapor evaporation. Small ingots of gold are kept in the thermal evaporator crucible. They are evaporated to be deposited on the membrane. Then it should be possible to turn the membrane to the other side so that the other side of Nafion membrane is also coated uniformly with gold. One may also hang it in a chamber so that both sides of the membrane are deposited with gold simultaneously. A better approach for electrode fabrication would have been sputtering of gold target onto the membrane. But thermal evaporators are easily available in small laboratories for microfabrication. Hence this method is suggested. Once the membrane is ready then it is cut into a circular membrane (diameter 2.5 mm) and acts as a micropump diaphragm. Copper tape with adhesive on one side is used to cut copper electrodes to be placed as annular rings on both surfaces of the membrane as shown in Figure 11.1a for the actuator assembly. The nozzle diffuser-based structure was fabricated in polymethyl methacrylate (PMMA) substrate (1 mm thickness) with 100 μm depth. For this laser cutting of PMMA is suggested. Now the chamber area for the micropump is ready. Polypropylene sheet acts as the micropump diaphragm. Hence the chamber is covered with polypropylene sheet except for the chamber area as shown in Figure 11.1a. This chamber area was covered by the gold coated Nafion 115 membrane sandwiched between two copper sheet rings connected to external actuation source. The top PMMA sheet has viewport where laser beam of the laser Doppler vibrometer (LDV) instrument can be focused. Two inlet outlet ports reach

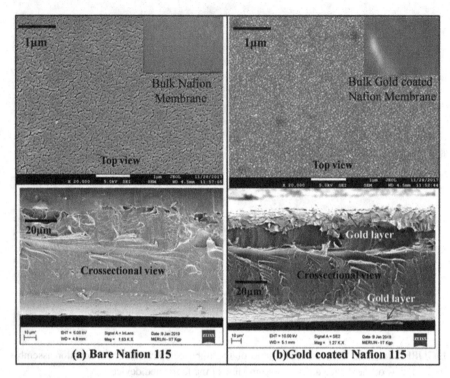

FIGURE 11.2 Scanning electron micrograph of top view and cross-sectional view of (a) bare Nafion 115 membrane and (b) gold coated Nafion membrane with pictures of respective bulk membranes in inset. The cross-sectional view shows the layer of gold layer deposited on both sides. Roughness inside profile is from the scissor cut membrane.

the backside of the bottom PMMA sheet and tubings are connected securely to them. Figure 11.1b and c shows the top and bottom view of the fabricated micropump.

Scanning electron microscopy-based (SEM) analysis is a way to determine the quality of gold coating in comparison to the uncoated Nafion membrane. Figure 11.2 shows the scanning electron micrographs of bare Nafion 115 and gold-coated Nafion 115 membrane. Figure 11.2a shows cracks in commercial Nafion 115 membrane which might have been introduced by the process-based stresses. These cracks, while aiding in membrane bending, have negative consequences like water leakage and consumption of more energy than membranes with less cracks. Figure 11.2b shows the uniform gold coating of Nafion membrane produced by physical vapor evaporation.

The frequency response of the Nafion 115 membrane is studied to determine its resonant frequency (Figure 11.3) on application of a 3 V square wave. In air, the resonant frequency for the Nafion 115 membrane is found to be around 150 Hz. The frequency is around the value shown in earlier works on IPMC-based cantilever structures where the natural frequency of Nafion like IPMC is around 100 Hz [33,34]. It shows that for applied 3 V, 0.5 Hz square wave, the dominant frequency

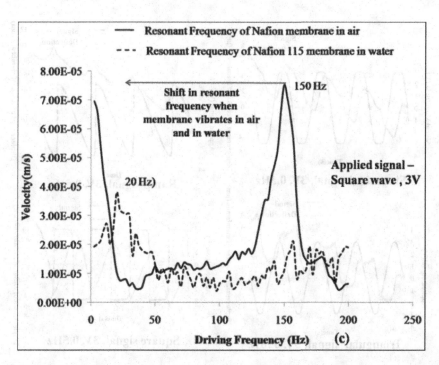

FIGURE 11.3 Frequency response of Nafion membrane with resonant frequencies shown for measurements in air and water.

response is around 0.5 Hz in air (applied signal's driving frequency) which shows that the membrane obeys the input frequency. It is seen that in the presence of DI water the resonant frequency shifts to 20 Hz. This can be attributed to membrane oscillation in the presence of DI water which is denser than air and viscous forces are exerted on it. It corresponds to an additional mass having additional inertial force being pushed by the membrane. Hence the resonant frequency for IPMC membrane gets shifted to lower frequencies when operating in DI water [35].

The Nafion membrane displays different responses for different types of electrical actuations. They are shown in Figure 11.4a. We note that a square wave input generates maximum deflection in the membrane. A quick response is observed in the case of a square wave and ramp signal, but the deformation gradually decreases due to membrane relaxation with time [36,37] for triangular and sinusoidal signals. Noting the better performance of the membrane on application of a square wave, this signal is used henceforth in the electrical characterizations.

The laser beam from an LDV was used to scan the entire length of the membrane while deflecting in air on application of 3 V, 0.5 Hz square wave. This deflection was compared with the simulation-based results and shown in Figure 11.5. The centre point witnesses the maximum deflection. The actual displacement is much less than the simulated data since the data was obtained using models which did not take air

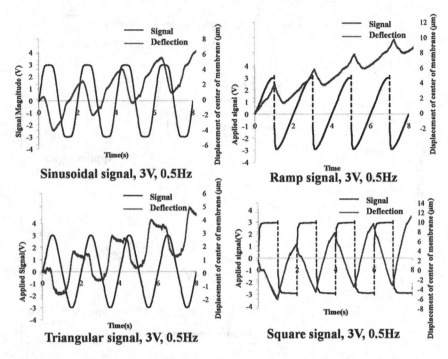

FIGURE 11.4   Response of Nafion 115 membrane for different type of actuating signals.

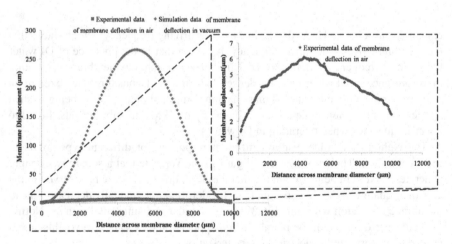

FIGURE 11.5   Graph showing comparison of simulation studies-based IPMC membrane deflection in vacuum and experimentally determined Nafion 115 membrane deflection in air for 3 V, 0.5 Hz sine wave.

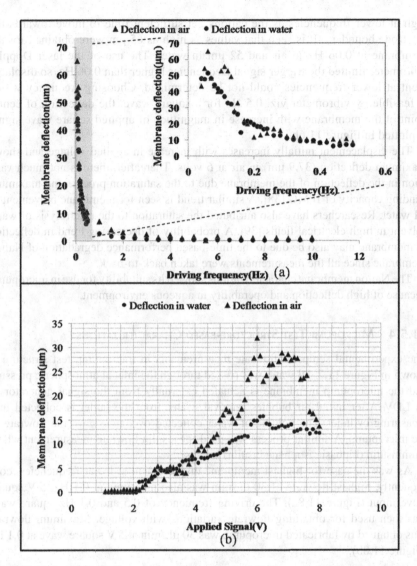

FIGURE 11.6 (a) Deflection of the pump membrane with increasing driving frequency with a 3 V square wave and (b) deflection of the IPMC membrane with different actuation voltages for a square wave 0.5 Hz. Maximum deflection is obtained for 6 V, 0.5 Hz square wave signal.

dampening into account. A detailed simulation model for IPMC membrane analysis is required which can incorporate damping by viscous medium as well.

Variation of IPMC membrane deflection with driving frequency for a square 3 V signal both in air as well as DI water is shown in Figure 11.6a. Frequency plays an important role in the IPMC membrane characterizations. The deflection remains lower for DI water at all values because of higher viscosity of DI water than air, but in both cases, it increases as frequency decreases and is particularly high at lower frequencies as the cations get sufficient time to move towards the electrode boundary. It is seen that values as high as 65 μm were obtained for the membrane at 0.06 Hz in air and 52 μm in water. The use of the laser Doppler vibrometer limited the trigger signal frequencies higher than 0.06 Hz, so displacement at lower frequencies could not be determined. Choosing frequency which is feasible by vibrometer viz. 0.5 Hz for a square wave, the deflection of centre point of the membrane with increase in magnitude of applied square wave signal is plotted in Figure 11.6b.

The displacement initially increases with increase in applied voltage and shows maximum deflection (37.9 μm) in air at 6 volts. Thereafter, there is not much variation in the deflection of the membrane due to the saturation process and maximum bending capacity of IPMC [38]. A similar trend is seen for membrane movement in DI water. Researchers have also attributed the saturation to the electrolysis of water solvent in high electrical fields [39]. A probability of decreasing trend in deflection of membrane may also be due to the time-based performance degradation of Nafion membrane since all the measurements were taken back-to-back.

The Nafion membrane characterizations showed its suitability for use in micropump because of high deflection and operability in aqueous environment.

11.5.4 MICROPUMP TEST STRUCTURE-BASED CHARACTERIZATIONS

The experimental setup for flowrate measurements in micropump test structure is shown in Figure 11.7. DI water was passed through the inlet at atmospheric pressure and the micropump membrane is actuated externally from the signal generator of an LDV. After initial air bubbles removal, water from the outlet is collected in a measuring cylinder. The volume of water collected over time gave the flowrate of the micropump. A similar process was used for insulin flowrate measurement where Huminsulin (biphasic isophane insulin) was used.

As was the case for highest membrane displacement at lower frequencies, consequently, highest flowrate of 30 μL/min was also measured at 0.1 Hz, 5 V square wave input (Figure 11.8a). The driving frequency of 0.5 and 0.1 Hz square wave was then used for obtaining flowrate variations with voltage. Maximum flowrate thus obtained by fabricated micropump was 30 μL/min at 5 V square wave at 0.1 Hz (Figure 11.8b).

11.5.5 SUMMARY

In this example, experimental studies were carried out to check the feasibility of using Nafion 115 membrane as micropump actuator diaphragm. Highest membrane deflection of 65 μm and maximum flowrate of 30 μL/min for Di water was found for 5–6 V, 0.5 Hz square wave signal. Hence this micropump presents an opportunity to be explored further and is promising candidate for drug delivery applications where the flowrate can be tuned by varying the frequency.

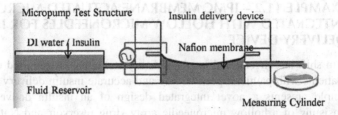

FIGURE 11.7 flowrate test characterization setup for Nafion membrane in micropump test structure.

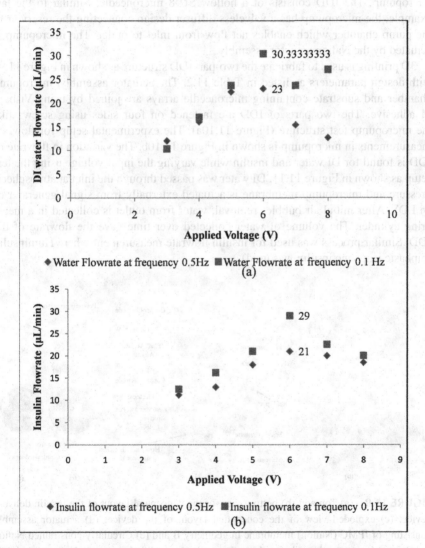

FIGURE 11.8 Plot of (a) DI water and (b) insulin flowrate variation with applied voltage when the Nafion membrane is used in test structure. Highest DI water and insulin flowrates are achieved for 6 V, 0.1 Hz square wave.

11.6   EXAMPLE 11.2 – IPMC MEMBRANE ACTUATED MICROPUMP INTEGRATED WITH HOLLOW MICRONEEDLES FOR INSULIN DELIVERY DEVICE

A paradigm shift in conventional insulin delivery techniques is required to provide diabetic patients with painless, precise, safe and accurate insulin delivery solutions. This example presents a novel integrated design of an insulin delivery device (IDD) consisting of a hollow microneedle array, drug reservoir and Nafion membrane actuation-based micropump. The proposed schematic is shown in Figure 11.9. It mainly consists of two MEMS-based components, namely microneedle and micropump. The IDD consists of a hollow SU-8 microneedle. Similar to the last example, the micropump has a valveless diffuser design connecting the reservoir to the pump chamber which enables net flow from inlet to outlet. The micropump is actuated by the Nafion actuator assembly.

3D printing is used to fabricate the two-part IDD structure as shown in Figure 11.9c with design parameters as listed in Table 11.2. The actuator assembly, micropump chamber and substrate containing microneedle arrays are joined by using Vishay-M adhesive. The two parts of IDD are fastened on four sides using screws like the micropump test structure (Figure 11.10a). The experimental setup for flowrate measurements in micropump is shown in Figure 11.10b. The variation in flowrate of IDD is found for DI water and insulin while varying the input voltage using the test setup as shown in Figure 11.11. DI water was passed through the inlet at atmospheric pressure and micropump membrane is actuated externally from signal generator of an LDV. After initial air bubbles removal, water from outlet is collected in a measuring cylinder. The volume of water collected over time gave the flowrate of the IDD. Similar process was used for insulin flowrate measurement where Huminsulin (biphasic isophane insulin was used).

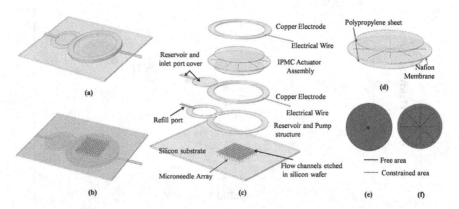

FIGURE 11.9   (a) Top and (b) bottom view of the conceptual layout of the insulin delivery device, (c) exploded view of the conceptual layout of the device, (d) actuator assembly consisting of IPMC (Nafion) membrane in geometry B and (e) Circularly constrained Nafion membrane (Geometry A) and (f) schematic of cuts in modified geometry Nafion membrane (geometry B).

TABLE 11.2
Parameters for Insulin Delivery Device (IDD)

S/n	Parameter	Value
1	IPMC material	Nafion 115
2	IPMC membrane width	10 mm
3	IPMC membrane thickness	0.127 mm
4	Gold electrode thickness	0.008 mm
5	Micropump chamber material	PMMA
6	PMMA sheet thickness	1 mm
7	Polymer sheet thickness	100 μm
8	Micropump chamber depth in PMMA	100 μm
9	Micropump chamber diameter	9 mm
10	Rubber gaskets thickness	1.5 mm
11	Diffuser length	1700 μm
12	Diffuser inlet width	300 μm
13	Diffuser outlet width	750 μm
14	Divergence angle	10°
15	Distance between diffuser and outlet / inlet	8300 μm
16	Microneedle array(10X10) area	9 mm
17	Reservoir diameter	5 mm
18	Reservoir, micropump material	PLA (Polylactic Acid)

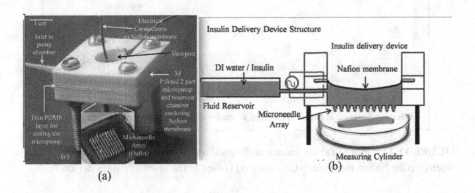

FIGURE 11.10 PLA-based Insulin delivery device fabricated by 3D printing with SU-8 hollow microneedle array in the inset and flowrate test characterization setup for Nafion membrane in insulin delivery device structure.

When Nafion membrane is assembled in the IDD, the outbound fluid must pass through 100 conduits (10X10 array) (microneedle lumen) instead of a single outlet. Hence it is expected that the fluid flow shall reduce owing to the flow resistance caused at each channel in the microneedle lumen. Figure 11.11 shows the flowrate plot obtained while varying the input signal for both DI water and insulin. It is seen

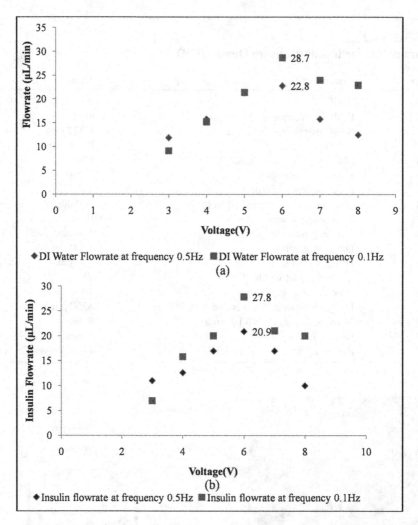

FIGURE 11.11   Plot of (a) DI water and (b) insulin flowrate variation with applied voltage when circular Nafion membrane (Geometry A) is used in the insulin delivery device.

that in agreement with our expectation on flow resistance, the flowrate reduces at all voltage values considered.

11.7   EXAMPLE 11.3 – MODIFIED GEOMETRY IPMC-BASED IDD FOR HIGH FLOWRATE

We saw in the last example that the IPMC membrane actuator assembly had Nafion membrane sandwiched between copper annular electrodes. This single piece diaphragm is clamped at all edges. This conventional design only allows up and down movement of the membrane. A design problem arises here of how to get increased

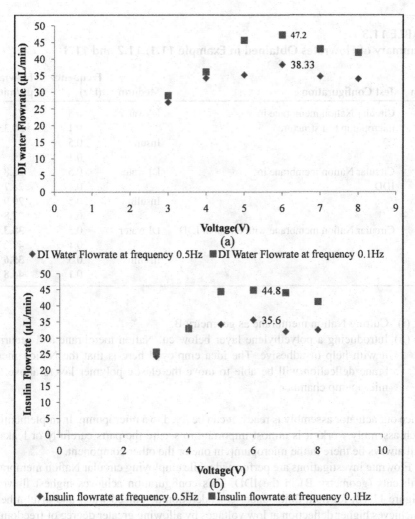

FIGURE 11.12 Plot of (a) DI water and (b) insulin flowrate variation with applied voltage when Nafion membrane is used in the modified circular geometry (geometry B).

flowrate with modification in the diaphragm. An approach could be to introduce a modified geometry to the circular membrane. It could be done by introducing cuts in the membrane. Here we consider an example where we cut the IPMC membrane and make them an assembly of eight cut structures instead of one for a circular membrane. This proposed geometry is shown in Figure 11.9f. We call this geometry as geometry B while naming the earlier one used in example 11.2 as A. This geometry allows more deflection as each part can act as a cantilever and the deflection would be greatly enhanced in case of an IPMC membrane. So, an assembly of eight such cantilever structures instead of one circular structure shall yield higher flowrate of device. Now the additional modification in the actuator assembly consists of following two steps:

TABLE 11.3
Summary of Flowrates Obtained in Example 11.1, 11.2 and 11.3

S/n	Test Configuration	Medium	Frequency (Hz)	Flowrate μL/min
1	Circular Nafion membrane in micropump test structure	DI water	0.5	23
			0.1	30.33
		Insulin	0.5	21
			0.1	29
2	Circular Nafion membrane in IDD	DI water	0.5	22.8
			0.1	28.7
		Insulin	0.5	20.9
			0.1	27.8
3	Circular Nafion membrane with 8 cuts in IDD	DI water	0.5	**38.33**
			0.1	**47.2**
		Insulin	0.5	**35.6**
			0.1	**44.8**

(i) Cutting Nafion membrane as geometry B.
(ii) Introducing a polyethylene layer below cut Nafion membrane and securing it with help of adhesive. The idea employed here is that the Nafion membrane deflection will be able to move the elastic polymer layers above the micropump chamber.

Once our actuator assembly is ready, it can be fixed to a micropump. In implementing such assembly work, it is utmost important to secure the parts carefully or leakage will always be there in the micropump in one or the other component.

Flowrate investigations are performed while employing circular Nafion membrane with cuts (geometry B) in the IDD. This configuration achieves highest flowrate (Figure 11.12) for both fluids under consideration deionized water and insulin here. It achieves higher deflection at low voltages by allowing greater degree of freedom in agreement with our previous work [38]. With everything remaining the same, these flowrates go high up to 47.2 μL/min for DI water and 44.8 μL/min for insulin at 6 V square wave at 0.1 Hz. The flowrate characterization results from all three examples have been summarized in Table 11.3.

11.8 CONCLUSION

This chapter covered the journey of different kinds of MEMS-based devices covered in previous chapters, to be integrated into a single drug delivery device. This integration was addressed while addressing a pressing drug delivery device challenge which is insulin delivery. Diabetes has become a dominant disease, causing deaths the world over and such devices are quite in demand. The fabrication steps and subsequently, feasibility studies have been performed on the device after taking into account the development of a customized test bench for such devices. The examples discussed in this work have been taken from the earlier works done in this field and elaborated for readers convenience.

REFERENCES

1. P.K. Das, A.B.M.T. Hasan, "Mechanical micropumps and their applications: A review", *AIP Conf. Proc.* 1851(1), 2017.
2. Y.N, Wang, L.M. Fu, "Micropumps and biomedical applications–A review", *Microelectron. Eng.* 195:121–138, 2018.
3. H. Genslar, R. Sheybani, P.Y. Li, R. Lo, E. Ming, "An implantable MEMS micropump system for drug delivery in small animals", *Biomed. Microdevices.* 14(3):483–496, 2012.
4. K. Siebal, L. Scholer, H. Schafer, N. Bohm, "A programmable planar electro-osmotic micropump for lab-on–chip applications", *J. Micromech. Microeng.* 18(2):025008, 2008.
5. R.R.V. Chemitiganti, C.W. Spellman. "Management of progressive type 2 diabetes: role of insulin therapy", *Osteopathic Med. Primary Care* 3(1):5, 2009.
6. S.E. Inzucchi, D.R Matthews, "Management of hyperglycemia in type 2 diabetes, 2015: A patient-centered approach. Update to a position statement of the American diabetes association and the European association for the study of diabetes," *Diabetes Care*, 38:140–149, Jan. 2015.
7. D.W. Dunstan, B.A. Kingwell, R. Larsen, G.N. Healy, E. Cerin, M.T. Hamilton, J. Shaw, D.A. Bertovic, P.Z. Zimmet, J. Salmon, N. Owen, "Breaking up prolonged sitting reduces postprandial glucose and insulin responses," *Diabetes Care.*35(5):976–983, 2012.
8. C. Joshitha, B.S. Sreeja, S. Radha, "A review on micropumps for drug delivery system," in *Proc. Int. Conf. Wireless Commun. Signal Process. Netw. (WiSPNET)*, Chennai, India, Mar., 186–190, 2017. DOI: 10.1109/WiSPNET.2017.8299745
9. H.A.E. Benson, A.C. Watkinson, *Topical and Transdermal Drug Delivery Principles and Practice.* Hoboken, NJ, USA: Wiley, 6–7, 2012.
10. V. Jain, "Management of type 1 diabetes in children and adolescents," *Indian J. Pediatr.* 81:170–177, Feb. 2014.
11. R. Aathira, V. Jain, "Advances in management of type 1 diabetes mellitus," *World J. Diabetes.* 5(5):689–696, 2014.
12. R. Mishra 1, T. K. Maiti, T. K. Bhattacharyya, "Design and scalable fabrication of hollow SU-8 microneedles for transdermal drug delivery," *IEEE Sensors J.* 18(14):5635–5644, Jul. 2018.
13. N.T. Nguyen, X. Huang T.K. Chuan, "MEMS-Micropumps: A Review", *J. Fluids Eng.* 124:384–392, 2012. DOI:10.1115/1.1459075
14. D.J. Laser, J.G. Santiago, "A review of micropumps", *J. Micromech. Microeng.* 14:R35–R64, 2004.
15. J.D. Zahn, "Micropump applications in Bio-MEMS", in W. Wang, W. Soper, S. A., *Bio-MEMS–Technologies and Applications* 4, 142–176, CRC Press, 2007.
16. M.W. Ashraf, S. Tayabba, N. Afzulpurkar, "Micro Electromechanical Systems (MEMS) Based Microfluidic devices for Biomedical Applications," *Int. J. Mol. Sci.* 12:3648–3704, 2011.
17. A.K. Yetisen, M.S. Akram, C.R. Lowe, "Paper-based microfluidic point-of-care diagnostic devices", *Lab. Chip.* 13:2210–2251, 2013.
18. G.S. Jeong, J. Oh, S.B. Kim, M.R. Dokmeci, H. Bae, S.H Lee, A. Khademhosseini, "Siphon-driven microfluidic passive pump with a yarn flow resistance controller". *Lab. Chip.* 14:4213–4219, 2014.

19. S. Singh, N. Kumar, D. George, A.K. Sen. "Analytical modeling, simulations and experimental studies of a PZT actuated planar valveless PDMS micropump", *Sens. Actuators A: Phys.* 225:81–94, 2015.

20. L. Hines, K. Petersen, G. Z. Lum, M. Sitti, "Soft actuators for small-scale robotics", *Advan. Mater.* 29(43):1603483, 2017.

21. D. Rus, M.T. Tolley, "Design, fabrication and control of soft robots", *Nature* 521(7553):467–475, 2015.

22. C. Jo, D. Pugal, I. Oh, K.J. Kim, K. Asaka, "Recent advances in ionic polymer–metal composite actuators and their modeling and applications", *Prog. Polym. Sci.* 38(7):1037–1066, 2013.

23. M. Porfiri, "An electromechanical model for sensing and actuation of ionic polymer metal composites", *Smart Mater. Struct.* 18(1):015016, 16, 2008.

24. Y. Bahramzadeh, M. Shahinpoor, "A review of ionic polymeric soft actuators and sensors", *Soft Robot.* 1(1):38–52, 2014.

25. R. Tiwari, E. Garcia, "The state of understanding of ionic polymer metal composite architecture: A review", *Smart Mater. Struct.* 20(8), 083001, 2011.

26. P.R. Resnick, W.G. Grot, E.I. Du Pont de Nemours and Company, "Method and Apparatus for Electrolysis of Alkali or Alkaline Earth Metal Halide", United States Patent, US 4113585, 1978.

27. C. Heitner-Wirguin, "Recent advances in perfluorinated ionomer membranes: Structure, properties and applications," *J. Membr. Sci.* 120:1–33, 1996.

28. K.A. Mauritz, R.B. Moore, "State of understanding of Nafion", *Chem. Rev.* 104(10):4535–4586, 2004.

29. https://en.wikipedia.org/wiki/Nafion accessed on 19.09.2018.

30. A. Dutta, R. Mishra, S. Biswas, J. Manna, R. Dhar, T.K. Bhattacahryya, Fabrication and characterization of Lithium doped nafion membrane and hollow glassy carbon microneedle for micropump based drug delivery, *MicroTAS*, 2020.

31. M.L. Guilly, M. Uchida, M. Taya, "Nafion-based smart membraneas an actuator array," *Proc. SPIE.* 4695:78–84, 2002. DOI: 10.1117/12.475151

32. Nafion Membranes: Fuelcellstore. Accessed: Jan. 16, 19. [Online]. Available: www.fuelcellstore.com/membranes/nafion

33. S. Lee, K.J. Kim, "Design of IPMC actuator-driven valveless micropump and its flow rate estimation at low Reynolds numbers", *Smart Mater. Struct.* 15(4):1103–1109, 2006.

34. T.T. Nguyen, S.N. Goo, V. Nguyen, Y. Yoo, S. Park, "Design, fabrication, and experimental characterization of a flap valve IPMC micropump with a flexibly supported diaphragm", *Sens. Actuators A: Phys.* 141:640–648, 2008.

35. P. Brunetto, L. Fortuna, S. Graziani, S. Strazzeri, "A model of ionic polymer–metal composite actuators in underwater operations", *Smart Mater. Struct.* 17(025029):12, 2008. DOI:10.1088/0964-1726/17/2/025029,

36. M. Shahinpoor, K.J. Kim., "Ionic polymer–metal composites: III. Modeling and simulation as biomimetic sensors, actuators, transducers, and artificial muscles", *Smart Mater. Struct.* 13(6):1362–1388, 2004.

37. M. Yu, H. Shen, Z. Dai, "Manufacture and performance of ionic polymer-metal composites", *J. Bionic Eng.* 4:143–149, 2007.

38. D. Kim, K.J. Kim, Y. Tak, D. Pugal, I.S. Park, "A self-oscillating electroactive polymer," *Appl. Phys. Letter.* 90(184104):1–3, 2007.

39. P.J. Costa Branco, J.A. Dente, "Derivation of a continuum model and its electric equivalent-circuit representation for ionic polymer–metal composite (IPMC) electromechanics", *Smart Mater. Struct.* 15(2):378–392, 2006.

Index

Printed in the United States
by Baker & Taylor Publisher Services

ted in the United States
& Taylor Publisher Services